200 Puzzling Physics Problems is a ....asp of
the laws of physics by applying tl .... and to
problems that yield more easily to ....ethods
and complex mathematics. The pr........ ....y from
classical (i.e. non-quantum) physics, but are no easier for that. For the most
part, these problems are intriguingly posed in accessible non-technical language.
This requires the student to select the right framework in which to analyse the
situation and to make decisions about which branches of physics are involved.
The general level of sophistication needed to tackle most of the 200 problems is
that of the exceptional school student, the good undergraduate or the competent
graduate student. The book should be valuable to undergraduates preparing for
'general physics' papers, either on their own or in classes or seminars designed
for this purpose. It is even hoped that some physics professors will find the more
difficult questions challenging. By contrast, the mathematical demands made are
minimal, and do not go beyond elementary calculus. This intriguing book of
physics problems should prove not only instructive and challenging, but also fun.

PETER GNÄDIG graduated as a physicist from Roland Eötvös University (ELTE) in Budapest
in 1971 and received his PhD in theoretical particle physics there in 1980. Currently, he is
a researcher (in high energy physics) and a lecturer in the Department of Atomic Physics
at ELTE. Since 1985 he has been one of the leaders of the Hungarian Olympic team taking
part in the International Physics Olympiad. He is also the Physics Editor of *KÖMAL*, the
100-year-old Hungarian *Mathematical and Physical Journal of Secondary Schools*, which
publishes several challenging physics problems each month, as well as one of the organisers
of the formidable Hungarian Physics competition (the Eötvös Competition). Professor
Gnädig has written textbooks on the theory of distributions and the use of vector-calculus
in physics.

GYULA HONYEK graduated as a physicist from Eötvös University (ELTE) in Budapest in
1975 and finished his PhD studies there in 1977, after which he stayed on as a researcher and
lecturer in the Department of General Physics. In 1984, following a two-year postgraduate
course, he was awarded a teacher's degree in physics, and in 1985 transferred to the
teacher training school at ELTE. His current post is as mentor and teacher at Radnóti
Grammar School, Budapest. Since 1986 he has been one of the leaders and selectors
of the Hungarian team taking part in the International Physics Olympiad. He is also a
member of the editorial board of *KÖMAL*. As a co-author of a physics textbook series
for Hungarian secondary schools, he has wide experience of teaching physics at all levels.

KEN RILEY read mathematics at the University of Cambridge and proceeded to a PhD
there in theoretical and experimental nuclear physics. He became a research associate in
elementary particle physics at Brookhaven, and then, having taken up a lectureship at the
Cavendish Laboratory, Cambridge, continued this research at the Rutherford Laboratory
and Stanford; in particular, he was involved in the discovery of a number of the early
baryonic resonances. As well as being Senior Tutor at Clare College, where he has
taught physics and mathematics for over 30 years, he has served on many committees
concerned with teaching and examining of these subjects at all levels of tertiary and
undergraduate education. He is also one of the authors of *Mathematical Methods for
Physics and Engineering* (Cambridge University Press).

200 Puzzling Physics Problems

P. Gnädig
Eötvös University, Budapest

G. Honyek
Radnóti Grammar School, Budapest

K. F. Riley
Cavendish Laboratory, Fellow of Clare College, Cambridge

CAMBRIDGE
UNIVERSITY PRESS

PUBLISHED BY THE PRESS SYNDICATE OF THE UNIVERSITY OF CAMBRIDGE
The Pitt Building, Trumpington Street, Cambridge, United Kingdom

CAMBRIDGE UNIVERSITY PRESS
The Edinburgh Building, Cambridge CB2 2RU, UK
40 West 20th Street, New York, NY 10011-4211, USA
477 Williamstown Road, Port Melbourne, 3207, Australia
Ruiz de Alarcón 13, 28014 Madrid, Spain
Dock House, The Waterfront, Cape Town 8001, South Africa

http://www.cambridge.org

First Published 2001
First South Asian Edition 2002
Reprinted 2003, 2005, 2008, 2009, 2012

Printed in India at Replika Press Pvt. Ltd., Kundli 131 028

Typeface Monotype Times 10/13 pt *System* LaT$_{\mathrm{E}}$X   [UPH]

A catalogue record for this book is available from the British Library

Library of Congress Cataloguing in Publication Data

Gnädig Péter, 1947-
200 Puzzling Physics Problems / P. Gnädig, G. Honyek, K. F. Riley.
p. cm.
Includes bibliographical reference and index.
ISBN 0521 77306 7 - ISBN 0 521 77480 2 (pb.)
1. Physics-Problems, exercises, etc. I. Title: Two hundred puzzling physics problems.
II. Honeyk, G. (Gyula), 1951-   III. Riley, K. F. (Kenneth Franklin), 1936-   IV. Title.
QC32.G52 2001
530'.076-dc21   00-53005   CIP

ISBN 978 0 521 54078 0

Special edition for sale in South Asia only, not for export elsewhere.

Contents

Preface   *page* vii

How to use this book   x

Thematic order of the problems   xi

Physical constants   xiii

Problems   1

Hints   50

Solutions   69

Preface

In our experience, an understanding of the laws of physics is best acquired by applying them to practical problems. Frequently, however, the problems appearing in textbooks can be solved only through long, complex calculations, which tend to be mechanical and boring, and often drudgery for students. Sometimes, even the best of these students, the ones who possess all the necessary skills, may feel that such problems are not attractive enough to them, and the tedious calculations involved do not allow their 'creativity' (genius?) to shine through.

This little book aims to demonstrate that not all physics problems are like that, and we hope that you will be intrigued by questions such as:

- How is the length of the day related to the side of the road on which traffic travels?
- Why are Fosbury floppers more successful than Western rollers?
- How far below ground must the water cavity that feeds Old Faithful be?
- How high could the tallest mountain on Mars be?
- What is the shape of the water bell in an ornamental fountain?
- How does the way a pencil falls when stood on its point depend upon friction?
- Would a motionless string reaching into the sky be evidence for UFOs?
- How does a positron move when dropped in a Faraday cage?
- What would be the high-jump record on the Moon?
- Why are nocturnal insects fatally attracted to light sources?
- How much brighter is sunlight than moonlight?
- How quickly does a fire hose unroll?

- How do you arrange two magnets so that the mutual couples they experience are not equal and opposite?
- How long would it take to defrost an 8-tonne Siberian mammoth?
- What perils face titanium-eating little green men who devour their own planet?
- What is the direction of the electric field due to an uniformly charged rod?
- What is the catch in an energy-generating capacitor?
- What is the equivalent resistance of an n-dimensional cube of resistors?
- What factors determine the period of a sand-glass egg timer?
- How does a unipolar dynamo work?
- How 'deep' is an electron lying in a box?

These, and some 180 others, are problems that can be solved elegantly by an appropriate choice of variables or coordinates, an unusual way of thinking, or some 'clever' idea or analogy. When such an inspiration or eureka moment occurs, the solution often follows after only a few lines of calculation or brief mental reasoning, and the student feels justifiably pleased with him- or herself.

Logic in itself is not sufficient. Nobody can guess these creative approaches without knowing and understanding the basic laws of physics. Accordingly, we would not encourage anybody to tackle these problems without first having studied the subject in some depth. Although successful solutions to the problems posed are clearly the principal goal, we should add that success is not to be measured by this alone. Whatever help you, the reader, may seek, and whatever stage you reach in the solution to a problem, it will hopefully bring you both enlightenment and delight. We are sure that some solutions will lead you to say 'how clever', others to say 'how nice', and yet others to say 'how obvious or heavy-handed'! Our aim is to show you as many useful 'tricks' as possible in order to enlarge your problem-solving arsenal. We wish you to use this book with delight and profit, and if you come across further similar 'puzzling' physics problems, we would ask you to share them with others (and send them to the authors).

The book contains 200 interesting problems collected by the authors over the course of many years. Some were invented by us, others are from the Hungarian 'Secondary School Mathematics and Physics Papers' which span more than 100 years. Problems and ideas from various Hungarian and international physics contests, as well as the Cambridge Colleges' entrance examination, have also been used, often after rewording. We have also been guided by the suggestions and remarks of our colleagues. It is impossible to determine the original authors of most of the physics problems appearing in the international 'ideas-market'. Nevertheless, some of the inventors of the most puzzling problems deserve our special thanks. They include Tibor Bíró, László Holics, Frederick Károlyházy, George Marx, Ervin Szegedi and István Varga. We thank them and the other people, known and unknown, who have authored, elaborated and improved upon 'puzzling' physics problems.

P.G.   G.H.

Budapest 2000

K.F.R.

Cambridge 2000

How to use this book

The following chapter contains the *problems*. They do not follow each other in any particular thematic order, but more or less in *order of difficulty*, or in groups requiring *similar methods of solution*. In any case, some of the problems could not be unambiguously labelled as belonging to, say, mechanics or thermodynamics or electromagnetics. Nature's secrets are not revealed according to the titles of the sections in a text book, but rather draw on ideas from various areas and usually in a complex manner. It is part of our task to find out what type of problem we are facing. However, *for information*, the reader can find a list of topics, and the problems that more or less belong to these topics, on the following page. Some problems are listed under more than one heading. The symbols and numerical values of the principal physical constants are then given, together with astronomical data and some properties of material.

The majority of the problems are not easy; some of them are definitely difficult. You, the reader, are naturally encouraged to try to solve the problems on your own and, obviously, if you do, you will get the greatest pleasure. If you are unable to achieve this, you should not give up, but turn to the relevant page of the short *hints* chapter. In most cases this will help, though it will not give the complete solution, and the details still have to be worked out. Once you have done this and want to check your result (or if you have completely given up and only want to see the *solution*), the last chapter should be consulted.

Problems whose solutions require similar reasoning usually follow each other. But if a particular problem relates to another elsewhere in this book, you will find a cross-reference in the relevant hint or solution. Those requiring especially difficult reasoning or mathematically complicated calculations are marked by one or two *asterisks*.

Some problems are included whose solutions raise further questions that are beyond the scope of this book. Points or issues worth further consideration are indicated at the end of the respective solutions, but the answers are not given.

Thematic order of the problems

Kinematics: 1, 3, 5, 36, 37, 38*, 40, 41, 64, 65*, 66, 84*, 86*.

Dynamics: 2, 7, 8, 12, 13, 24, 32*, 33, 34, 35, 37, 38*, 39*, 70*, 73*, 77, 78*, 79*, 80**, 82, 83, 85**, 90, 154*, 183*, 184*, 186*, 193*, 194.

Gravitation: 15, 16, 17, 18, 32*, 81**, 87, 88, 109, 110*, 111*, 112*, 116, 134*.

Mechanical energy: 6, 7, 17, 18, 32*, 51, 107.

Collisions: 20, 45, 46, 47, 48, 71, 72*, 93, 94, 144*, 194, 195.

Mechanics of rigid bodies: 39*, 42**, 58, 60*, 61**, 94, 95*, 96, 97*, 98, 99**.

Statics: 9, 10*, 11, 14*, 25, 26, 43, 44, 67, 68, 69*.

Ropes, chains: 4, 67, 81**, 100, 101*, 102**, 103*, 104*, 105**, 106*, 108**

Liquids, gases: 19, 27, 28, 49*, 50, 70*, 73*, 74, 75*, 91*, 115**, 143, 200.

Surface tension: 29, 62, 63, 129, 130*, 131*, 132**, 143, 199*.

Thermodynamics: 20, 21*, 133, 135**, 136, 145, 146*, 147, 148.

Phase transitions: 134*, 137*, 138, 140*, 141*.

Optics: 52, 53, 54, 55, 56, 125*, 126, 127, 128*.

Electrostatics: 41, 90, 91*, 92, 113*, 114, 117*, 118, 121, 122, 123*, 124*, 149, 150, 151*, 152, 155, 156, 157, 183*, 192*, 193*.

Magnetostatics: 89**, 119, 120**, 153*, 154*, 172, 186*.

Electric currents: 22, 23, 158, 159, 160*, 161, 162*, 163*, 164*, 165, 169, 170*, 172.

Electromagnetism: 30, 31, 166, 167, 168*, 171*, 173*, 174*, 175*, 176, 177, 178*, 179, 180, 181*, 182*, 184*, 185*, 186*, 187*.

Atoms and particles: 93, 188, 189*, 190*, 191, 194, 195, 196, 197*, 198*.

Dimensional analysis, scaling, estimations: 15, 57, 58, 59*, 76*, 77, 126, 139, 142, 185*, 199*.

*,** A single or double asterisk indicates those problems that require especially difficult reasoning or mathematically complicated calculations.

Physical constants

Gravitational constant, G	$6.673 \times 10^{-11} \, \mathrm{N\,m^2\,kg^{-2}}$
Speed of light (in vacuum), c	$2.998 \times 10^8 \, \mathrm{m\,s^{-1}}$
Elementary charge, e	$1.602 \times 10^{-19} \, \mathrm{C}$
Electron mass, m_e	$9.109 \times 10^{-31} \, \mathrm{kg}$
Proton mass, m_p	$1.673 \times 10^{-27} \, \mathrm{kg}$
Boltzmann constant, k	$1.381 \times 10^{-23} \, \mathrm{J\,K^{-1}}$
Planck constant, h	$6.626 \times 10^{-34} \, \mathrm{J\,s}$
Avogadro constant, N_A	$6.022 \times 10^{23} \, \mathrm{mol^{-1}}$
Gas constant, R	$8.315 \, \mathrm{J\,mol^{-1}\,K^{-1}}$
Permittivity of free space, ε_0	$8.854 \times 10^{-12} \, \mathrm{C\,V^{-1}\,m^{-1}}$
Coulomb constant, $k = 1/4\pi\varepsilon_0$	$8.987 \times 10^9 \, \mathrm{V\,m\,C^{-1}}$
Permeability of free space, μ_0	$4\pi \times 10^{-7} \, \mathrm{V\,s^2\,C^{-1}\,m^{-1}}$

Some astronomical data

Mean radius of the Earth, R	$6371 \, \mathrm{km}$
Sun–Earth distance (Astronomical Unit, AU)	$1.49 \times 10^8 \, \mathrm{km}$
Mean density of the Earth, ρ	$5520 \, \mathrm{kg\,m^{-3}}$
Free-fall acceleration at the Earth's surface, g	$9.81 \, \mathrm{m\,s^{-2}}$

Some physical properties

Surface tension of water, γ	$0.073 \, \mathrm{N\,m^{-1}}$
Heat of vaporisation of water, L	$2256 \, \mathrm{kJ\,kg^{-1}} = 40.6 \, \mathrm{kJ\,mol^{-1}}$
Tensile strength of steel, σ	$500\text{--}2000 \, \mathrm{MPa}$

Densities[a], $\rho \, (\mathrm{kg \, m^{-3}})$

Hydrogen	0.0899	Titanium	4510
Helium	0.1786	Iron	7860
Air	1.293	Mercury	13 550
Water (at 4 °C)	1000	Platinum	21 450

[a] Densities quoted in normal state.

Optical Refractive Indices[b], n

Water	1.33	Glass	1.5–1.8
Ice	1.31	Diamond	2.42

[b] At $\lambda = 590$ nm.

Problems

P1 Three small snails are each at a vertex of an equilateral triangle of side 60 cm. The first sets out towards the second, the second towards the third and the third towards the first, with a uniform speed of 5 cm min^{-1}. During their motion each of them always heads towards its respective target snail. How much time has elapsed, and what distance do the snails cover, before they meet? What is the equation of their paths? If the snails are considered as point-masses, how many times does each circle their ultimate meeting point?

P2 A small object is at rest on the edge of a horizontal table. It is pushed in such a way that it falls off the other side of the table, which is 1 m wide, after 2 s. Does the object have wheels?

P3 A boat can travel at a speed of 3 m s^{-1} on still water. A boatman wants to cross a river whilst covering the shortest possible distance. In what direction should he row with respect to the bank if the speed of the water is (i) 2 m s^{-1}, (ii) 4 m s^{-1}? Assume that the speed of the water is the same everywhere.

P4 A long, thin, pliable carpet is laid on the floor. One end of the carpet is bent back and then pulled backwards with constant unit velocity, just above the part of the carpet which is still at rest on the floor.

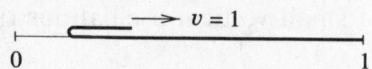

Find the speed of the centre of mass of the moving part. What is the minimum force needed to pull the moving part, if the carpet has unit length and unit mass?

P5   Four snails travel in uniform, rectilinear motion on a very large plane surface. The directions of their paths are random, (but not parallel, i.e. any two snails could meet), but no more than two snail paths can cross at any one point. Five of the $(4 \times 3)/2 = 6$ possible encounters have already occurred. Can we state with certainty that the sixth encounter will also occur?

P6   Two 20-g flatworms climb over a very thin wall, 10 cm high. One of the worms is 20 cm long, the other is wider and only 10 cm long. Which of them has done more work against gravity when half of it is over the top of the wall? What is the ratio of the amounts of work done by the two worms?

P7   A man of height $h_0 = 2$ m is bungee jumping from a platform situated a height $h = 25$ m above a lake. One end of an elastic rope is attached to his foot and the other end is fixed to the platform. He starts falling from rest in a vertical position.

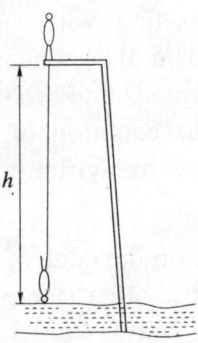

The length and elastic properties of the rope are chosen so that his speed will have been reduced to zero at the instant when his head reaches the surface of the water. Ultimately the jumper is hanging from the rope, with his head 8 m above the water.

(i) Find the unstretched length of the rope.
(ii) Find the maximum speed and acceleration achieved during the jump.

P8   An iceberg is in the form of an upright regular pyramid of which 10 m shows above the water surface. Ignoring any induced motion of the water, find the period of small vertical oscillations of the berg. The density of ice is 900 kg m^{-3}.

P9   The suspension springs of all four wheels of a car are identical. By how much does the body of the car (considered rigid) rise above each of the wheels when its right front wheel is parked on an 8-cm-high pavement? Does the result change when the car is parked with both right wheels on the pavement? Does the result depend on the number and positions of the people sitting in the car?

P10* In Victor Hugo's novel *les Misérables*, the main character Jean Valjean, an escaped prisoner, was noted for his ability to climb up the corner formed by the intersection of two vertical perpendicular walls. Find the minimum force with which he had to push on the walls whilst climbing. What is the minimum coefficient of static friction required for him to be able to perform such a feat?

P11 A sphere, made of two non-identical homogeneous hemispheres stuck together, is placed on a plane inclined at an angle of 30° to the horizontal. Can the sphere remain in equilibrium on the inclined plane?

P12 A small, elastic ball is dropped vertically onto a long plane inclined at an angle α to the horizontal. Is it true that the distances between consecutive bouncing points grow as in an arithmetic progression? Assume that collisions are perfectly elastic and that air resistance can be neglected.

P13 A small hamster is put into a circular wheel-cage, which has a frictionless central pivot. A horizontal platform is fixed to the wheel below the pivot. Initially, the hamster is at rest at one end of the platform.

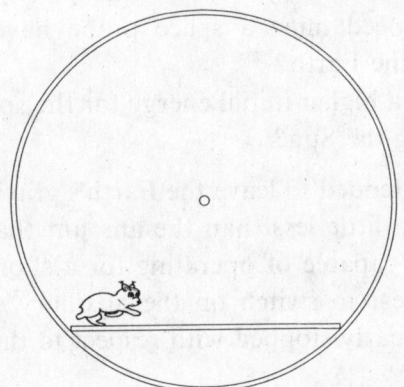

When the platform is released the hamster starts running, but, because of the hamster's motion, the platform and wheel remain *stationary*. Determine how the hamster moves.

P14* A bicycle is supported so that it is prevented from falling sideways but can move forwards or backwards; its pedals are in their highest and lowest positions. A student crouches beside the bicycle and applies a horizontal force, directed towards the back wheel, to the lower pedal.

(i) Which way does the bicycle move?

    (ii) Does the chain-wheel rotate in the same or opposite sense as the rear wheel?

    (iii) Which way does the lower pedal move relative to the ground?

P15   If the solar system were proportionally reduced so that the average distance between the Sun and the Earth were 1 m, how long would a year be? Take the density of matter to be unchanged.

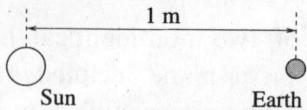

P16   If the mass of each of the members of a binary star were the same as that of the Sun, and their distance apart were equal to the Sun–Earth distance, what would be their period of revolution?

P17   (i) What is the minimum launch speed required to put a satellite into a circular orbit?

(ii) How many times higher is the energy required to launch a satellite into a polar orbit than that necessary to put it into an Equatorial one?

(iii) What initial speed must a space probe have if it is to leave the gravitational field of the Earth?

(iv) Which requires a higher initial energy for the space probe – leaving the solar system or hitting the Sun?

P18   A rocket is intended to leave the Earth's gravitational field. The fuel in its main engine is a little less than the amount that is necessary, and an auxiliary engine, only capable of operating for a short time, has to be used as well. When is it best to switch on the auxiliary engine: at take-off, or when the rocket has nearly stopped with respect to the Earth, or does it not matter?

P19   A steel ball with a volume of 1 cm^3 is sinking at a speed of 1 cm s^{-1} in a closed jar filled with honey. What is the momentum of the honey if its density is 2 g cm^{-3}?

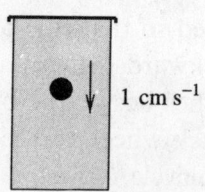

P20   A gas of temperature T is enclosed in a container whose walls are (initially) at temperature T_1. Does the gas exert a higher pressure on the walls of the container when $T_1 < T$ or when $T_1 > T$?

P21*   Consider two identical iron spheres, one of which lies on a thermally insulating plate, whilst the other hangs from an insulating thread.

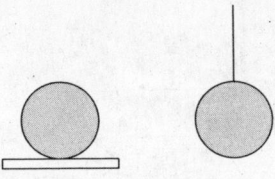

Equal amounts of heat are given to the two spheres. Which will have the higher temperature?

P22   Two (non-physics) students, A and B, living in neighbouring college rooms, decided to economise by connecting their ceiling lights in series. They agreed that each would install a 100-W bulb in their own rooms and that they would pay equal shares of the electricity bill. However, both decided to try to get better lighting at the other's expense; A installed a 200-W bulb and B installed a 50-W bulb. Which student subsequently failed the end-of-term examinations?

P23   If a battery of voltage V is connected across terminals I of the black box shown in the figure, a voltmeter connected to terminals II gives a reading of $V/2$; while if the battery is connected to terminals II, a voltmeter across terminals I reads V.

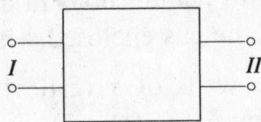

The black box contains only passive circuit elements. What are they?

P24   A bucket of water is suspended from a fixed point by a rope. The bucket is set in motion and the system swings as a pendulum. However, the bucket leaks and the water slowly flows out of the bottom of it. How does the period of the swinging motion change as the water is lost?

P25   An empty cylindrical beaker of mass 100 g, radius 30 mm and negligible wall thickness, has its centre of gravity 100 mm above its base. To what depth should it be filled with water so as to make it as stable as possible?

P26   Fish soup is prepared in a hemispherical copper bowl of diameter 40 cm. The bowl is placed into the water of a lake to cool down and floats with 10 cm of its depth immersed.

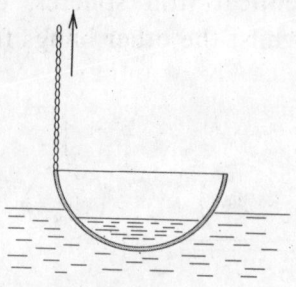

A point on the rim of the bowl is pulled upwards through 10 cm, by a chain fastened to it. Does water flow into the bowl?

P27   A circular hole of radius r at the bottom of an initially full water container is sealed by a ball of mass m and radius $R(>r)$. The depth of the water is now slowly reduced, and when it reaches a certain value, h_0, the ball rises out of the hole. Find h_0.

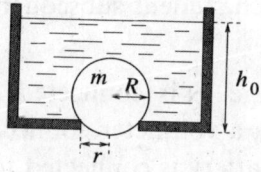

P28   Soap bubbles filled with helium float in air. Which has the greater mass – the wall of a bubble or the gas enclosed within it?

P29   Water which wets the walls of a vertical capillary tube rises to a height H within it. Three 'gallows', (a), (b) and (c), are made from the same tubing, and one end of each is placed into a large dish filled with water, as shown in the figure.

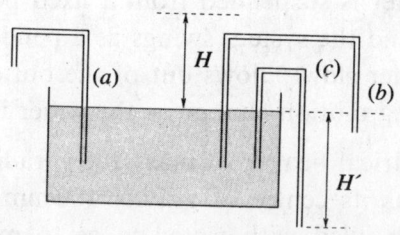

Does the water flow out at the other ends of the capillary tubes?

P30 A charged spherical capacitor slowly discharges as a result of the slight conductivity of the dielectric between its concentric plates. What are the magnitude and direction of the magnetic field caused by the resulting electric current?

P31 An electrically charged conducting sphere 'pulses' radially, i.e. its radius changes periodically with a fixed amplitude (*see figure*). The charges on its surface – acting as many dipole antennae – emit electromagnetic radiation. What is the net pattern of radiation from the sphere?

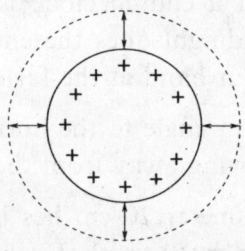

P32* How high would the male world-record holder jump (at an indoor competition!) on the Moon?

P33 A small steel ball *B* is at rest on the edge of a table of height 1 m. Another steel ball *A*, used as the bob of a metre-long simple pendulum, is released from rest with the pendulum suspension horizontal, and swings against *B* as shown in the figure. The masses of the balls are identical and the collision is elastic.

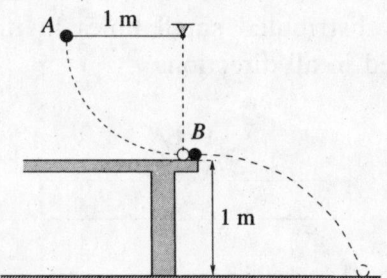

Considering the motion of *B* only up until the moment it first hits the ground:

  (i) Which ball is in motion for the longer time?
  (ii) Which ball covers the greater distance?

P34 A small bob is fixed to one end of a string of length 50 cm. As a consequence of the appropriate forced motion of the other end of the string, the bob moves in a vertical circle of radius 50 cm with a uniform speed of 3.0 m s^{-1}. Plot, at 15° intervals on the circular path, the trajectories of both ends of the string, indicating on each the points belonging together.

P35 A point P is located above an inclined plane. It is possible to reach the plane by sliding under gravity down a straight frictionless wire, joining P to some point P' on the plane. How should P' be chosen so as to minimise the time taken?

P36 The minute hand of a church clock is twice as long as the hour hand. At what time after midnight does the end of the minute hand move away from the end of the hour hand at the fastest rate?

P37 What is the maximum angle to the horizontal at which a stone can be thrown and always be moving away from the thrower?

P38[*] A tree-trunk of diameter 20 cm lies in a horizontal field. A lazy grasshopper wants to jump over the trunk. Find the minimum take-off speed of the grasshopper that will suffice. (Air resistance is negligible.)

P39[*] A straight uniform rigid hair lies on a smooth table; at each end of the hair sits a flea. Show that if the mass M of the hair is not too great relative to that m of each of the fleas, they can, by simultaneous jumps with the same speed and angle of take-off, exchange ends without colliding in mid-air.

P40 A fountain consists of a small hemispherical rose (sprayer) which lies on the surface of the water in a basin, as illustrated in the figure. The rose has many evenly distributed small holes in it, through which water spurts at the same speed in all directions.

What is the shape of the water 'bell' formed by the jets?

P41 A particle of mass m carries an electric charge Q and is subject to the combined action of gravity and a uniform horizontal electric field of strength E. It is projected with speed v in the vertical plane parallel to the field and at an angle θ to the horizontal. What is the maximum distance the particle can travel horizontally before it is next level with its starting point?

P42[**] A uniform rod of mass m and length ℓ is supported horizontally at its ends by my two forefingers. Whilst I am slowly bringing my fingers together to meet under the centre of the rod, it slides on either one or other of them. How much work do I have to do during the process if the coefficient of static friction is μ_{stat}, and that of kinetic friction is μ_{kin} ($\mu_{kin} \le \mu_{stat}$)?

P43 Four identical bricks are placed on top of each other at the edge of a table. Is it possible to slide them horizontally across each other in such a way that the projection of the topmost one is completely outside the table? What is the theoretical limit to the displacement of the topmost brick if the number of bricks is arbitrarily increased?

P44 A plate, bent at right angles along its centre line, is placed onto a horizontal fixed cylinder of radius R as shown in the figure.

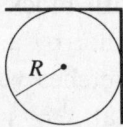

How large does the coefficient of static friction between the cylinder and the plate need to be if the plate is not to slip off the cylinder?

P45 Two elastic balls of masses m_1 and m_2 are placed on top of each other (with a small gap between them) and then dropped onto the ground. What is the ratio m_1/m_2, for which the upper ball ultimately receives the largest possible fraction of the total energy? What ratio of masses is necessary if the upper ball is to bounce as high as possible?

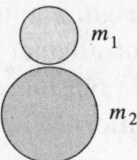

P46 An executive toy consists of three suspended steel balls of masses M, μ and m arranged in that order with their centres in a horizontal line. The ball of mass M is drawn aside in their common plane until its centre has been raised by h and is then released. If $M \ne m$ and all collisions are elastic, how must μ be chosen so that the ball of mass m rises to the greatest possible height? What is this height? (Neglect multiple collisions.)

P47 Two identical dumb-bells move towards each other on a horizontal air-cushioned table, as shown in the figure. Each can be considered as two point masses m joined by a weightless rod of length 2ℓ. Initially, they are not rotating. Describe the motion of the dumb-bells after their elastic collision. Plot the speeds of the centres of mass of the dumb-bells as a function of time.

P48   Two small identical smooth blocks A and B are free to slide on a frozen lake. They are joined together by a light elastic rope of length $\sqrt{2}L$ which has the property that itⁱ stretches very little when the rope becomes taut. At time $t = 0$, A is at rest at $x = y = 0$ and B is at $x = L$, $y = 0$ moving in the positive y-direction with speed V. Determine the positions and velocities of A and B at times (i) $t = 2L/V$ and (ii) $t = 100L/V$.

P49*   After a tap above an empty rectangular basin has been opened, the basin fills with water in a time T_1. After the tap has been closed, opening a plug-hole at the bottom of the basin empties it in a time T_2. What happens if both the tap and the plug-hole are open? What ratio of T_1/T_2 can cause the basin to overflow? As a specific case, let $T_1 = 3$ minutes and $T_2 = 2$ minutes.

P50   A cylindrical vessel of height h and radius a is two-thirds filled with liquid. It is rotated with constant angular velocity ω about its axis, which is vertical. Neglecting any surface tension effects, find an expression for the greatest angular velocity of rotation Ω for which the liquid does not spill over the edge of the vessel.

P51   Peter, who was standing by a racetrack, calculated that as one of the cars, in accelerating from rest to a speed of 100 km h^{-1}, used up x litres of fuel, it could increase its speed to 200 km h^{-1}, by using a further $3x$ litres of fuel. Peter, who has learned in physics that kinetic energy is proportional to the square of the speed, assumed that the energy content of the fuel was mainly converted into kinetic energy, i.e. he neglected air resistance and other types of friction.

A railway runs by the racetrack. Paul, who also knows some physics, saw the start of the race from the window of a train travelling at a speed of 100 km h^{-1} in the opposite direction to that of the car. He reasoned as follows: since the car's speed increased from 100 to $200\,\mathrm{km\,h^{-1}}$ during the first stage, when the car accelerates from 200 to $300\,\mathrm{km\,h^{-1}}$ in the second stage, it will need $(300^2 - 200^2)/(200^2 - 100^2)\,x = (5/3)x$ litres of fuel.

Who is right, Peter or Paul?

P52 The distance between a screen and a light source lined up on an optical bench is 120 cm. When a lens is moved along the line joining them, sharp images of the source can be obtained at two lens positions; the size (area) ratio of these two images is 1 : 9. What is the focal length of the lens? Which image is the brighter? Determine the ratio of the brightness values of the two images.

P53 A short-sighted person takes off his glasses and observes a fixed object through them, while moving the glasses away from his eyes. He is surprised to see that at first, the object looks smaller and smaller, but then becomes larger and larger. What is the reason for this?

P54 A glass prism whose cross-section is an isosceles triangle stands with its (horizontal) base in water; the angles that its two equal sides make with the base are each θ.

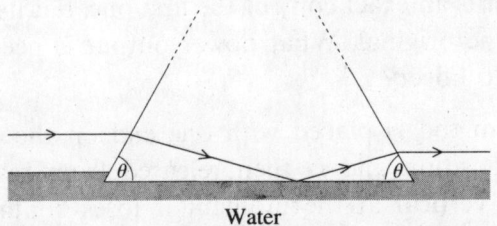

Water

An incident ray of light, above and parallel to the water surface and perpendicular to the prism's axis, is internally reflected at the glass–water interface and subsequently re-emerges into the air. Taking the refractive indices of glass and water to be $\frac{3}{2}$ and $\frac{4}{3}$, respectively, show that θ must be at least $25.9°$.

P55 A glass prism in the shape of a quarter-cylinder lies on a horizontal table. A uniform, horizontal light beam falls on its vertical plane surface, as shown in the figure.

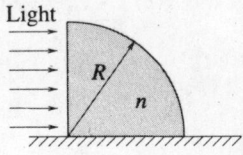

If the radius of the cylinder is $R = 5$ cm and the refractive index of the glass is $n = 1.5$, where, on the table beyond the cylinder, will a patch of light be found?

P56 How much brighter is sunlight than moonlight? The albedo (reflectivity) of the Moon is $\alpha = 0.07$.

P57 Annie and her very tall boyfriend Andy like jogging together. They notice that when running they move at more or less the same speed, but that Andy is always faster when they are walking. How can this difference between running and walking be explained using physical arguments?

P58 A simple pendulum and a homogeneous rod pivoted at its end are released from horizontal positions. What is the ratio of their periods of swing if their lengths are identical?

P59* A helicopter can hover when the power output of its engine is P. A second helicopter is an exact copy of the first one, but its linear dimensions are half those of the original. What power output is needed to enable this second helicopter to hover?

P60* A uniform rod is placed with one end on the edge of a table in a nearly vertical position and is then released from rest. Find the angle it makes with the vertical at the moment it loses contact with the table. Investigate the following two extreme cases:

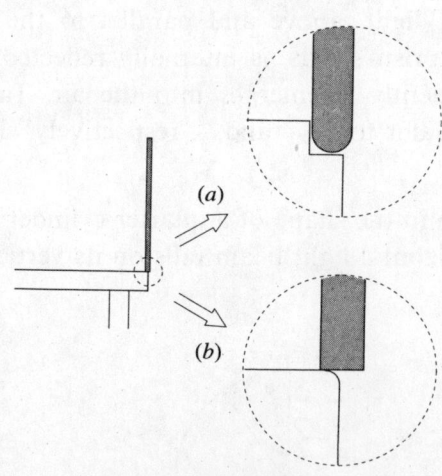

(i) The edge of the table is smooth (friction is negligible) but has a small, single-step groove as shown in figure (*a*).

(ii) The edge of the table is rough (friction is large) and very sharp, which means that the radius of curvature of the edge is much smaller than the flat end-face of the rod. Half of the end-face protrudes beyond the table edge (*see figure (b)*), with the result that when it is released from rest the rod 'pivots' about the edge. The rod is much longer than its diameter.

P61** A pencil is placed vertically on a table with its point downwards. It is then released and tumbles over. How does the direction in which the point moves, relative to that in which the pencil falls, depend upon the coefficient of friction? Will the pencil point lose contact with the table (other than when the 'shoulder' of the pencil ultimately comes into contact with the table)?

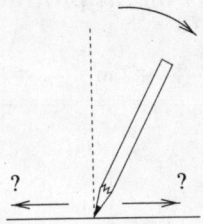

P62 Two soap bubbles of radii R_1 and R_2 are joined by a straw. Air goes from one bubble to the other (which one?) and a single bubble of radius R_3 is formed. What is the surface tension of the soap solution if the atmospheric pressure is p_0? Is measuring three such radii a suitable method for determining the surface tension of liquids?

P63 Water, which wets glass, is stuck between two parallel glass plates. The distance between the plates is d, and the diameter of the trapped water 'disc' is $D \gg d$.

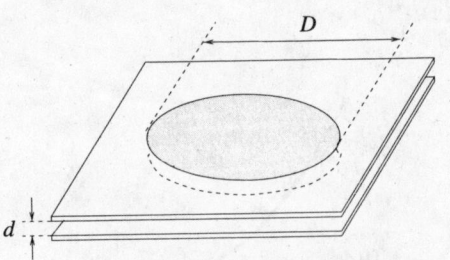

What is the force acting between the plates?

P64   A spider has fastened one end of a 'super-elastic' silk thread of length 1 m to a vertical wall. A small caterpillar is sitting somewhere on the thread.

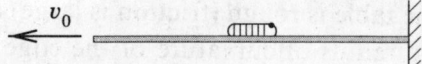

The hungry spider, whilst not moving from its original position, starts pulling in the other end of the thread with uniform speed, $v_0 = 1$ cm s^{-1}. Meanwhile, the caterpillar starts fleeing towards the wall with a uniform speed of 1 mm s^{-1} with respect to the moving thread. Will the caterpillar reach the wall?

P65   How does the solution to the previous problem change if the spider does not sit in one place, but moves (away from the wall) taking with it the end of the thread?

P66   Nails are driven horizontally into a vertically placed drawing-board. As shown in the figure, a small steel ball is dropped from point A and reaches point B by bouncing elastically on the protruding nails (which are not shown in the figure).

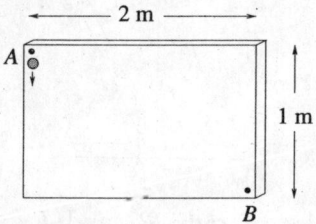

Is it possible to arrange the nails so that:

(i) The ball gets from point A to point B more quickly than if it had slid without friction down the shortest path, i.e. along the straight line AB?

(ii) The ball reaches point B in less than 0.4 s?

P67   One end of a rope is fixed to a vertical wall and the other end pulled by a horizontal force of 20 N. The shape of the flexible rope is shown in the figure. Find its mass.

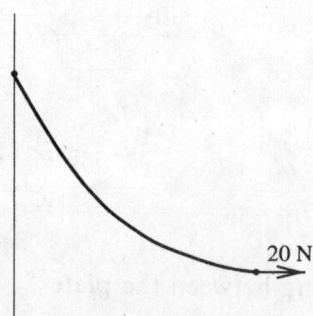

P68 Find the angle to which a pair of compasses should be opened in order to have the pivot as elevated as possible when the compasses are suspended from a string attached to one of the points, as shown in the figure. Assume that the lengths of the compass arms are equal.

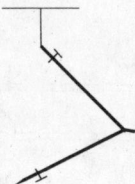

P69* Threads of lengths h_1, h_2 and h_3 are fastened to the vertices of a homogeneous triangular plate of weight W. The other ends of the threads are fastened to a common point, as shown in the figure.

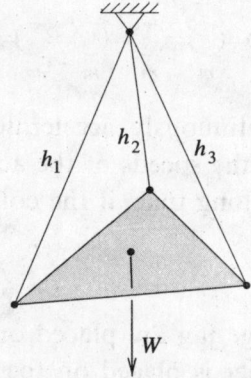

What is the tension in each thread, expressed in terms of the lengths of the threads and the weight of the plate?

P70* A tanker full of liquid is parked at rest on a horizontal road. The brake has not been applied, and it may be supposed that the tanker can move without friction.

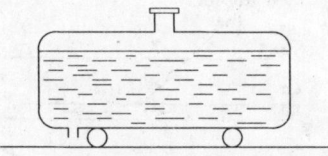

In which direction will the tanker move after the tap on the vertical outlet pipe, which is situated at the rear of the tanker, has been opened? Will the tanker continue to move in this direction?

P71   Two small beads slide without friction, one on each of two long, horizontal, parallel, fixed rods set a distance d apart. The masses of the beads are m and M, and they carry respective charges of q and Q. Initially, the larger mass M is at rest and the other one is far away approaching it at speed v_0.

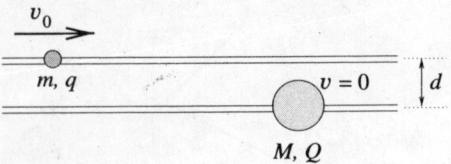

Describe the subsequent motion of the beads.

P72*   Beads of equal mass are strung at equal distances on a long, horizontal wire. The beads are initially at rest but can move without friction.

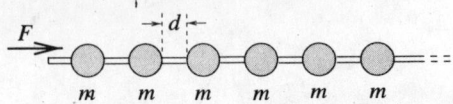

One of the beads is continuously accelerated (towards the right) by a constant force F. What are the speeds of the accelerated bead and the front of the 'shock wave', after a long time, if the collisions of the beads are:

(i) completely inelastic,

(ii) perfectly elastic?

P73*   A table and a large jug are placed on the platform of a weighing machine and a barrel of beer is placed on the table with its tap above the jug. Describe how the reading of the machine varies with time after the tap has been opened and the beer runs into the jug.

P74   A jet of water strikes a horizontal gutter of semicircular cross-section obliquely, as shown in the figure. The jet lies in the vertical plane that contains the centre-line of the gutter.

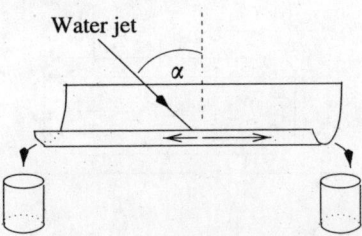

Water jet

Calculate the ratio of the quantities of water flowing out at the two ends of the gutter as a function of the angle of incidence α of the jet.

P75* An open-topped vertical tube of diameter D is filled with water up to a height h. The narrow bottom-end of the tube, of diameter d, is closed by a stop as shown in the figure.

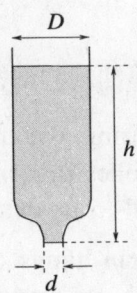

When the stop is removed, the water starts flowing out through the bottom orifice with approximate speed $v = \sqrt{2gh}$. However, this speed is reached by the liquid only after a certain time τ. Obtain an estimate of the order of magnitude of τ. What is the acceleration of the lowest layer of water at the moment when the stop is removed? Ignore viscous effects.

P76* Obtain a reasoned estimate of the time it takes for the sand to run down through an egg-timer. Use realistic data.

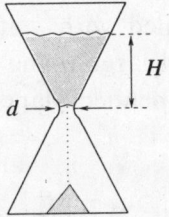

P77 A small bob joins two light unstretched, identical springs, anchored at their far ends and arranged along a straight line, as shown in the figure.

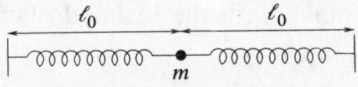

The bob is displaced in a direction perpendicular to the line of the springs by 1 cm and then released. The period of the ensuing vibration of the bob is 2 s. Find the period of the vibration if the bob were displaced by 2 cm before release. The unstretched length of the springs is $\ell_0 \gg 1$ cm, and gravity is to be ignored.

P78* One end of a light, weak spring, of unstretched length L and force constant k, is fixed to a pivot, and a body of mass m is attached to its other end. The spring is released from an unstretched, horizontal position, as in the figure.

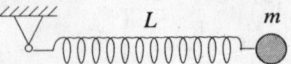

What is the length of the spring when it reaches a vertical position? (Describing a spring as weak implies that $mg \gg kL$, and that the tension in the spring is directly proportional to its extension at all times.)

P79* A heavy body of mass m hangs on a flexible thread in a railway carriage which moves at speed v_0 on a train-safety test track, as shown in the figure.

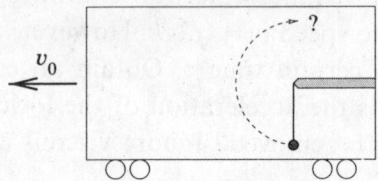

The carriage is brought to rest by a strong but uniform braking. Can the pendulum travel through 180°, so that the taut thread reaches the vertical?

P80** A glass partially filled with water is fastened to a wedge that slides, without friction, down a large plane inclined at an angle α as shown in the figure. The mass of the inclined plane is M, the combined mass of the wedge, the glass and the water is m.

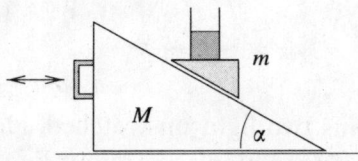

If there were no motion the water surface would be horizontal. What angle will it ultimately make with the inclined plane if

(i) the inclined plane is fixed,

(ii) the inclined plane can move freely in the horizontal direction?

Examine also the case in which $m \gg M$. What happens if the handle of the inclined plane is shaken in a periodic manner, but one that is such that it does not cause the wedge to rise off the plane?

P81** If someone found a motionless string reaching vertically up into the sky and hanging down nearly to the ground, should that person consider it as an evidence for UFOs, or could there be an 'Earthly' explanation in agreement with the well-known laws of physics? How long would the string need to be?

P82 There is a parabolic-shaped bridge across a river of width 100 m. The highest point of the bridge is 5 m above the level of the banks. A car of mass 1000 kg is crossing the bridge at a constant speed of 20 m s^{-1}.

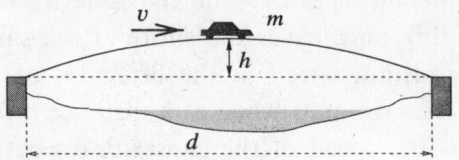

Using the notation indicated in the figure, find the force exerted on the bridge by the car when it is:

   (i) at the highest point of the bridge,

   (ii) three-quarters of the way across.

(Ignore air resistance and take g as 10 m s^{-2}.)

P83 A point mass of 0.5 kg moving with a constant speed of 5 m s^{-1} on an elliptical track experiences an outward force of 10 N when at either endpoint of the major axis and a similar force of 1.25 N at each end of the minor axis. How long are the axes of the ellipse?

P84* A boatman sets off from one bank of a straight, uniform canal for a mark directly opposite the starting point. The speed of the water flowing in the canal is v everywhere. The boatman rows steadily at such a rate that, were there no current, the boat's speed would also be v. He always sets the boat's course in the direction of the mark, but the water carries him downstream. Fortunately he never tires! How far downstream does the water carry the boat? What trajectory does it follow with respect to the bank?

P85** Two children stand on a large, sloping hillside that can be considered as a plane. The ground is just sufficiently icy that a child would fall and slide downhill with a uniform speed as the result of receiving even the slightest impulse.

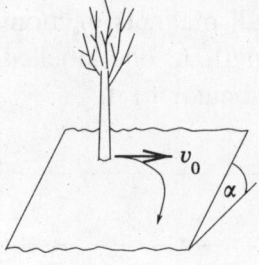

For fun, one of the children (leaning against a tree) pushes the other with a *horizontal* initial speed $v_0 = 1$ m s^{-1}. The latter slides down the slope with a velocity that changes in both magnitude and direction. What will be the child's final speed if air resistance is negligible and the frictional force is independent of the speed?

P86* Smugglers set off in a ship in a direction perpendicular to a straight shore and move at constant speed v. The coastguard's cutter is a distance a from the smugglers' ship and leaves the shore at the same time. The cutter always moves at a constant speed in the direction of the smugglers' ship and catches up with the criminals when at a distance a from the shore. How many times greater is the speed of the coastguard's cutter than that of the smugglers' ship?

P87 Point-masses of mass m are at rest at the corners of a regular n-gon, as illustrated in the figure for $\eta = 6$.

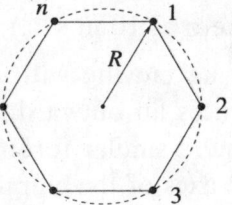

How does the system move if only gravitation acts between the bodies? How much time elapses before the bodies collide if $n = 2, 3$ and 10? Examine the limiting case when $n \gg 1$ and $m = M_0/n$, where M_0 is a given total mass.

P88 A rocket is launched from and returns to a spherical planet of radius R in such a way that its velocity vector on return is parallel to its launch vector. The angular separation at the centre of the planet between the launch and arrival points is θ. How long does the flight of the rocket take, if the period of a satellite flying around the planet just above its surface is T_0? What is the maximum distance of the rocket above the surface of the planet? Consider whether your analysis also applies to the limiting case of $\theta \to 0$.

P89** Two identical small magnets of moment μ are glued to opposite ends of a wooden rod of length L, one labelled C, parallel to the rod, and the other labelled D, perpendicular to it.

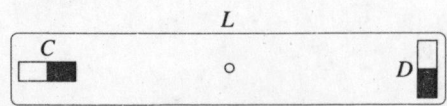

(i) Show that the couples that the magnets exert on each other are *not* equal and opposite.

(ii) Ignoring the Earth's magnetic field, explain quantitatively what would happen if the system were freely suspended at its centre of gravity.

P90 A point-like body of mass m and charge q is held above and close to a large metallic fixed plane and released when a distance d from it. How much time will it take for the body to reach the plane? Ignore gravity.

P91* A plastic ball, of diameter 1 cm and carrying a uniform charge of 10^{-8} C, is suspended by an insulating string with its lowest point 1 cm above a large container of brine (salted water). As a result, the surface of the water below the ball wells up a little.

How large is the rise in water level immediately below the ball? Ignore the effect of surface tension, and take the density of salted water to be 1000 kg m^{-3}.

P92 A point charge is at rest inside a thin metallic spherical shell, but is not at its centre. What is the force acting on the charge?

P93 Boron atoms of mass number $A = 10$ and a beam of unidentified particles, moving in opposite directions with the same (non-relativistic) speed, are made to collide inside an ion accelerator. The maximum scattering angle of the boron atoms is found to be 30°. What kind of atoms does the particle beam consist of?

P94 A billiard ball rolling without slipping hits an identical, stationary billiard ball in a head-on collision. Describe the motion of the balls after the collision. Prove that the final state does not depend on the coefficient of sliding friction between the balls and the billiard table. (Rolling friction is negligible.)

P95* A long slipway, inclined at an angle α to the horizontal, is fitted with many identical rollers, consecutive ones being a distance d apart. The rollers have horizontal axles and consist of rubber-covered solid steel cylinders each of mass m and radius r. Planks of mass M, and length much greater than d, are released at the top of the slipway.

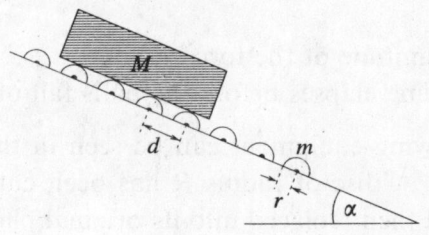

Find the terminal speed v_{max} of the planks. Ignore air resistance and friction at the pivots of the rollers.

P96 A tablecloth covers a horizontal table and a steel ball lies on top of it. The tablecloth is pulled from under the ball, and friction causes the ball to move and roll.

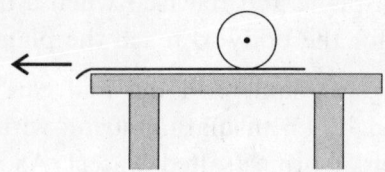

What is the ball's speed on the table when it reaches a state of rolling without slipping? (Assume that the table is so large that the ball does not fall off it.)

P97* If the law were changed so that traffic in Great Britain travelled on the right-hand side of the road (instead of on the left), would the length of the day increase, decrease, or be unaltered?

P98 In a physics stunt, two balls of equal density, and radii r and $R = 2r$, are placed with the centre of the larger one at the middle of a cart of mass $M = 6$ kg and length $L = 2$ m. The mass of the smaller ball is $m = 1$ kg. The balls are made to roll, without slipping, in such a way that the larger ball rests on the cart, and a straight line connecting their centres remains at a constant angle $\phi = 60°$ to the horizontal. The cart is pulled by a horizontal force in the direction shown in the figure.

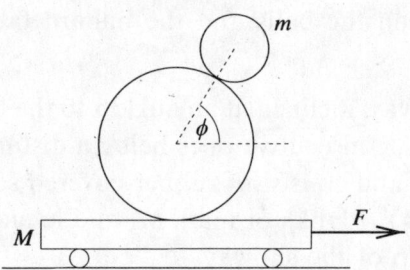

(i) Find the magnitude of the force F.
(ii) How much time elapses before the balls fall off the cart?

P99** The following equipment can be seen in the Science Museum in Canberra, Australia. A disc of radius R has been cut from the centre of a horizontal table, and then replaced into its original place mounted on a axle.

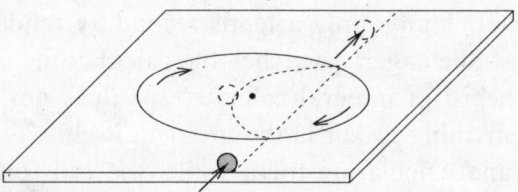

As illustrated in the figure, the disc is spun and a solid rubber ball is rolled onto the table. When it reaches the spinning disc, the ball leaves its straight-line course and follows a curve. On leaving the disc, it continues its *original* course, rolling without slipping, along a straight line. The final speed of the ball is the same as it was before it reached the disc.

What are the conservation principles underlying this motion?

P100 A thin ring of radius R is made of material of density ρ and Young's modulus E. It is spun in its own plane, about an axis through its centre, with angular velocity ω. Determine the amount (assumed small) by which its circumference increases.

P101* A light, inelastic thread is stretched round one-half of the circumference of a fixed cylinder as shown in the figure.

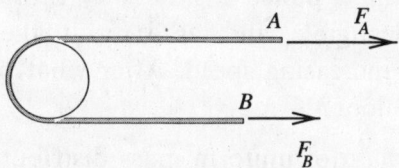

As a result of friction, the thread does not slip on the cylinder when the magnitudes of the forces acting on its ends fulfil the inequality

$$\frac{1}{2}F_A \le F_B \le 2F_A.$$

Determine the coefficient of friction between the thread and the cylinder.

P102** Charlie is a first-year student at university, studying integral calculus in mathematics. As an exercise, he has to determine the position C of the centre of mass of a semicircular arc which has radius R and a homogeneous mass distribution.

His younger sister, Jenny, only attends secondary school, but is studying rotation in physics. She eagerly watches the calculations of her brother, but as she has never heard of integral calculus, she does not understand much of it. The only clear thing to her is the problem itself.

After thinking and calculating for a while, she calls out: 'I have got the result, and I can determine not only the position of the centre of mass of a semicircle but also that of any part of a circle or any sector of it!'

How has she done it?

P103* A table of height 1 m has a hole in the middle of its surface. A thin, golden chain necklace, of length 1 m, is placed loosely coiled close to the hole, as shown in the figure.

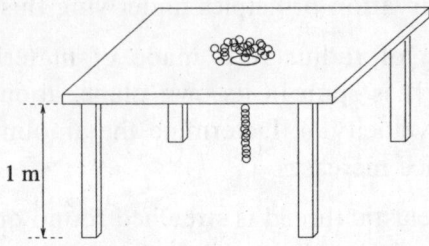

One end of the chain is pulled a little way through the hole and then released. Friction is negligible, and, as a result, the chain runs smoothly through the hole with increasing speed. After what times will the two ends of the chain reach the floor?

P104* A flexible chain of uniform mass distribution is wrapped tightly round two cylinders so that its form is that of a stadium running-track, i.e. it consists of two semicircles joined by two straight sections. The cylinders are made to rotate and cause the chain to move with speed v.

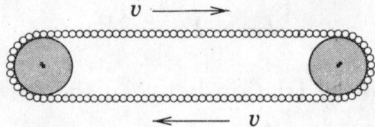

For some reason, the chain suddenly slips off the cylinders and falls vertically. How does the shape of the chain vary during the fall?

According to Steve, it takes a circular shape because of the centrifugal force. Bob accepts this point, but he considers that the initially 'elliptical' chain will be deformed beyond the circular by this effect and become a vertical ellipse with its new major axis at right angles to the original one. He expects that this process will repeat itself and that the chain shape will cycle between the two 'ellipses'. Frank guesses that the chain retains its original shape, but he cannot give any reasons for his guess. Who is right – or are they perhaps all wrong?

P105** A heavy, flexible, inelastic chain of length L is placed almost symmetrically onto a light pulley which can rotate about a fixed axle, as shown in the figure.

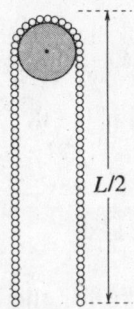

What will the speed of the chain be when it leaves the pulley?

P106* A long, heavy, flexible rope with mass ρ per unit length is stretched by a constant force F. A sudden movement causes a circular loop to form at one end of the rope. In a manner similar to that in which transverse waves propagate, the loop runs (rolls) along the rope with speed c as shown in the figure.

(i) Calculate the speed c of the loop.
(ii) Determine the energy, momentum and angular momentum carried by a loop of angular frequency ω. What is the relationship between these quantities?

P107 Sand falls vertically at a rate of 50 kg s^{-1} onto a horizontal conveyor belt moving at a speed of 1 m s^{-1}, as shown in the figure.

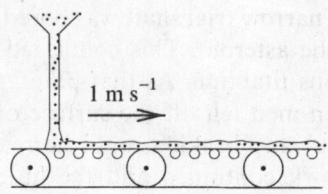

What is the minimum power output of the engine which drives the belt? How is the work done by the engine accounted for?

P108**   A fire hose of mass M and length L is coiled into a roll of radius R ($R \ll L$). The hose is sent rolling across level ground with initial speed v_0 (angular velocity v_0/R), while the free end of the hose is held at a fixed point on the ground. The hose unrolls and becomes straight.

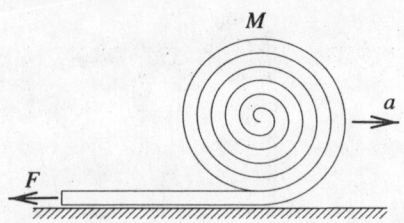

(i) How much time does it take for the hose to completely unroll?
(ii) The speed of the roll continually increases and its acceleration a is clearly a vector pointing in the same direction as its velocity. On the other hand, the vector resultant of the horizontal external forces (frictional force plus the restraining force at the fixed end of the hose) points in the opposite direction. How are these two facts consistent with Newton's second law?

(To simplify the analysis, suppose the initial kinetic energy of the roll to be much higher than its potential energy ($v_0 \gg \sqrt{gR}$), thus allowing the effect of gravity to be neglected. Assume further that the hose can be considered as arbitrarily flexible, and that the work necessary for its deformation, air resistance and rolling resistance can all be neglected.)

P109   Where is gravitational acceleration greater, on the surface of the Earth, or 100 km underground? Take the Earth as spherically symmetrical. The average density of the Earth is 5500 kg m^{-3}, and that of its crust is 3000 kg m^{-3}. (The depth of the crust may be assumed to be at least 100 km.)

P110*   The Examining Institute for Cosmic Accidents (EXINCA) sent the following short report to one of its experts:

> A spaceship of titanium-devouring little green people has found a perfectly spherical asteroid. A narrow trial shaft was bored from point A on its surface to the centre O of the asteroid. This confirmed that the whole asteroid is made of homogeneous titanium. At that point, an accident occurred when one of the little green men fell off the surface of the asteroid into the trial shaft. He fell, without any braking, until he reached O, where he died on impact. However, work continued and the little green men started secret

excavation of the titanium, in the course of which they formed a spherical cavity of diameter AO inside the asteroid, as illustrated in the figure.

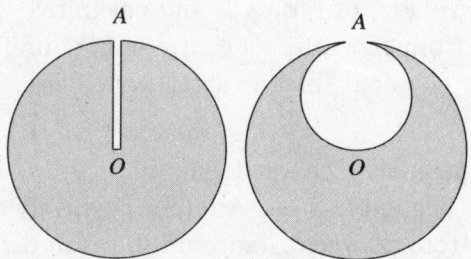

Then a second accident occurred – another little green man similarly fell from point A to point O, and died.

EXINCA asked the expert to calculate the ratio of the impact speeds and the ratio of the times taken to fall from A to O by the two unfortunate little men. What figures did the expert give in her reply?

P111* The titanium-devouring little green people of the previous problem continued their excavating. As a result of their environmentally destructive activity, half of the asteroid was soon used up and, as shown in the figure, only a regular hemisphere remained. The excavated material was carried away from the asteroid.

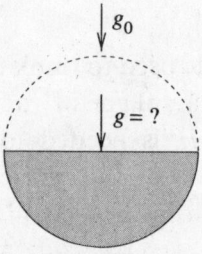

What is the gravitational acceleration at the centre of the circular face of the remaining hemisphere if the gravitational acceleration at the surface of the original (spherical) asteroid was $g_0 = 9.81$ cm s^{-2}?

P112* The little green titanium-devouring people found another titanium asteroid with a radius of 10 km and a homogeneous mass distribution. They started to excavate and to convey the material of the asteroid to the surface. The excavation of the metal was effected by boring shafts along a strip 1 m wide round the equator of the asteroid until they had cut the asteroid completely in two. Then the accident happened; the props separating the two hemispheres broke and the asteroid collapsed.

The experts from EXINCA need to calculate the *total* force exerted on the props just before they collapsed. Please help them.

P113*   A metal sphere, of radius R and cut in two along a plane whose minimum distance from the sphere's centre is h, is uniformly charged by a total electric charge Q. What force is necessary to hold the two parts of the sphere together?

P114   A small positively charged ball of mass m is suspended by an insulating thread of negligible mass. Another positively charged small ball is moved very slowly from a large distance until it is in the original position of the first ball. As a result, the first ball rises by h. How much work has been done?

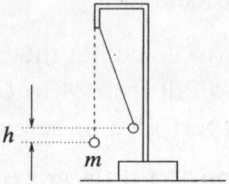

P115**   Hydrogen gas is stored at high pressure in a small, spherical container. The gas is introduced into a light balloon and its pressure becomes equal to the external atmospheric pressure. Is it possible that the balloon could lift the container in its final state? Assume that the temperature of the gas remains constant.

P116   In olden times, people used to think that the Earth was flat. Imagine that the Earth is indeed not a sphere of radius R, but an infinite plate of thickness H. What value of H is needed to allow the same gravitational acceleration to be experienced as on the surface of the actual Earth? (Assume that the Earth's density is uniform and equal in the two models.)

P117*   Electrical charges are evenly distributed along a long, thin insulating rod AB.

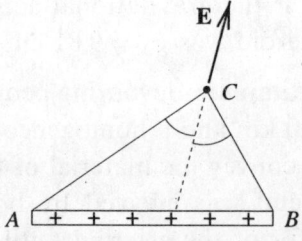

Show that at an arbitrary point C (*see figure*), the electric field due to the rod points in the direction of the bisector of angle ACB.

P118 Using the result of the previous problem, determine the direction and magnitude of the electric field in a plane which is perpendicular to a long, charged rod, and contains one of the rod's endpoints.

P119 At the beginning of nineteenth century the magnetic field of wires carrying currents was the focus of investigations in physics, both experimentally and theoretically. A particularly interesting case is that of a very long wire, carrying a constant current I, which has been bent into the form of a 'V', with opening angle 2θ.

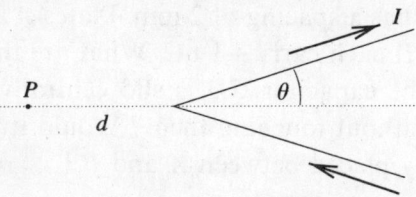

According to Ampère's computations, the magnitude B of the magnetic field at a point P lying outside the 'V', but on its axis of symmetry and at a distance d from its vertex, is proportional to $\tan(\theta/2)$. However, for the same situation, Biot and Savart suggested that the magnetic field at P might be proportional to θ. In fact they attempted to decide between the two possibilities by measuring the oscillation period of a magnetic needle as a function of the 'V' opening angle. However, for a range of θ values, the predicted differences were too small to be measured.

  (i) Which formula might be correct?
  (ii) Find the proportionality factor in this formula and guess the most likely factor appearing in the other one.

P120** A direct current flows in a solenoid of length L and radius R, ($L \gg R$), producing a magnetic field of magnitude B_0 inside the solenoid.

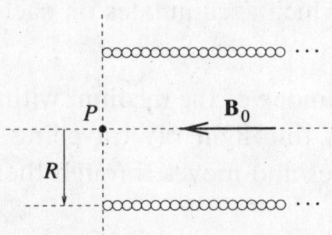

  (i) What is the strength of the magnetic field at the end of the coil, i.e. at the point P shown in the figure?

(ii) What is the magnetic flux at the end of the coil, i.e. through a virtual disc of radius R centred on P?

(iii) Sketch the magnetic field lines in the vicinity of P.

P121 The inner surfaces of two close parallel *insulating* plates are each given a uniform charge of $+Q$. What force is required to hold the plates together? Sketch the electric field lines of this arrangement.

P122 Two parallel plate capacitors differ only in the spacing between their (very thin) plates; one, AB, has a spacing of 5 mm and a capacitance of 20 pF, the other, CD, has a spacing of 2 mm. Plates A and C carry charges of $+1$ nC, whilst B and D each carry -1 nC. What are the potential differences V_{AB} and V_{CD} after the capacitor CD is slid centrally between and parallel to the plates of AB without touching them? Would it make any difference if CD were not centrally placed between A and B?

P123* The distance between the plates of a plane capacitor is d and the area of each plate is A. As shown in the figure, both plates of the capacitor are earthed and a small body carrying charge Q is placed between them, at a distance x from one plate.

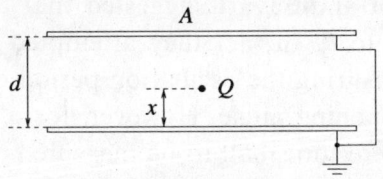

What charge will accumulate on each plate?

P124* A point-like electric dipole is placed between the earthed plates of the plane capacitor discussed in the previous problem. Its dipole momentum vector **p** is perpendicular to the plates and the distances of the dipole from the plates are x and $d - x$, respectively.

How does the charge which accumulates on each of the plates depend on x? (Ignore edge effects.)

P125* The refractive index of the medium within a certain region, $x > 0$, $y > 0$, changes with y. A thin light ray travelling in the x-direction strikes the medium at right angles and moves through the medium along a circular arc.

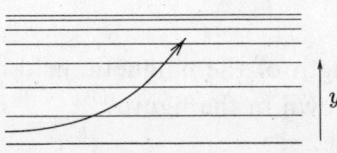

How does the refractive index depend on y? What is the maximum possible angular size of the arc?

P126 A compact disc (CD) contains approximately 650 MB of information. Estimate the size of one bit on a CD using an ordinary ruler. Confirm your estimate using a laser beam. Can you suggest the shape of one unit of information?

P127 When a particular line spectrum is examined using a diffraction grating of 300 lines mm^{-1} with the light at normal incidence, it is found that a line at 24.46° contains both red (640–750 nm) and blue/violet (360–490 nm) components. Are there any other angles at which the same thing would be observed?

P128* A parallel, thin, monochromatic laser beam falls on a diffraction grating at normal incidence. How does the interference pattern it produces on a viewing screen change if the grating is rotated through an angle $\phi < 90°$ around an axis, which is

(i) parallel to the lines of the grating; or

(ii) perpendicular to the lines of the grating?

P129 Two floating objects are attracted to each other as the result of surface tension effects, irrespective of whether they are floating on water or on mercury. Explain why this is so.

P130* Water in a clean aquarium forms a meniscus, as illustrated in the figure.

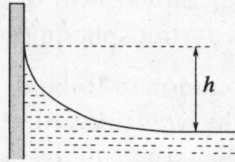

Calculate the difference in height h between the centre and the edge of the meniscus. The surface tension of water is $\gamma = 0.073$ N m^{-1}.

P131* Is it possible to have a (spherical) drop of water that could evaporate without taking up heat or losing internal (thermal) energy?

P132** Small liquid drops of various sizes are in a closed container, to whose walls the liquid does not adhere. Over a sufficiently long time, the size of the smallest drops is found to decrease whilst that of the larger ones increases, until finally only one large drop remains in the container. What is the explanation for this phenomenon?

P133 A horizontal frictionless piston, of negligible mass and heat capacity, divides a vertical insulated cylinder into two halves. Each half of the cylinder contains 1 mole of air at standard temperature and pressure p_0.

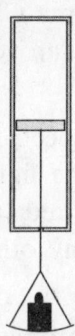

A load of weight W is now suspended from the piston, as shown in the figure. It pulls the piston down and comes to rest after a few oscillations. How large a volume does the compressed air in the lower part of the cylinder ultimately occupy if W is very large?

P134* How high could the tallest mountain on Earth be? And on Mars?

P135** The sealed lower half of a straight glass tube, of height 152 cm, is filled with air. The top half contains mercury and the top of the tube is left open. The air is slowly heated. How much heat has been transferred to the air by the time all the mercury has been pushed out of the tube?

Make a plot showing how the molar heat of the enclosed air changes with its volume during the process. (Atmospheric pressure is 760 mm Hg.)

P136 Vulcanism is very common in Iceland, but glaciers cover 11 per cent of its surface area. This is why volcanic eruptions quite often occur under glaciers, as one did in October 1996 under Vatnajokull, Europe's largest glacier. At the site of the eruption the glacier was 500 m thick and more or less smooth and flat. After a day's activity the visible sign of the eruption was a deep crater-like depression on the surface of the ice cap, in the form of a upside-down cone with a depth of 100 m and a diameter of 1 km. Explain the formation of the depression. What would have been found under the ice crater at this time? Try to predict the subsequent events.

P137* The most famous geyser in Yellowstone National Park is Old Faithful. This geyser can be considered as a large underground cavity with a narrow flue leading to the surface.

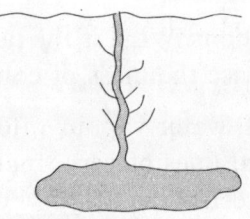

The surrounding earth is warm as a result of residual volcanic activity and boils the water in the cavity. After coming to the boil, the water in the flue is expelled, and approximately 44 tons of steam leave the geyser in 4 minutes. After the eruption, underground springs refill the cavity and the flue to ground level in 20–30 minutes, and the process then repeats itself. An eruption occurs every 90 minutes.

Geological experiments show that the underground temperature in this area increases by 1 °C for each metre of depth. Determine the minimum distance below the surface at which the cavity is situated. If the cavity is assumed to be located at this minimum depth, what is its volume?

P138 The air above a large lake is at −2 °C, whilst the water of the lake is at 0 °C. Assuming that only thermal conduction is important, and using relevant data selected from that given below, estimate how long it would take for a layer of ice 10 cm thick to form on the lake's surface.

Data:

Thermal conductivity of water, $\lambda_w = 0.56 \, \text{W} \, \text{m}^{-1} \, \text{K}^{-1}$
Thermal conductivity of ice, $\lambda_i = 2.3 \, \text{W} \, \text{m}^{-1} \, \text{K}^{-1}$
Specific latent heat of fusion of ice, $L_i = 3.3 \times 10^5 \, \text{J} \, \text{kg}^{-1}$
Density of water, $\rho_w = 1000 \, \text{kg} \, \text{m}^{-3}$
Density of ice, $\rho_i = 920 \, \text{kg} \, \text{m}^{-3}$

P139 If it takes two days to defrost a frozen 5-kg turkey, estimate how long it would take to defrost an 8-tonne Siberian mammoth.

P140* A 0.6-kg block of ice at −10 °C is placed into a closed empty 1 m³ container, also at a temperature of −10 °C. The temperature of the container is then increased to 100 °C. How much greater is the heat required than that necessary to raise the empty container alone to that temperature?

P141* A strong-walled container is half-filled with water. The other half contains air, initially at standard temperature and pressure. The container is closed and slowly heated. When does the water in the container start boiling? In what state(s) does the water exist, as the temperature rises?

P142 Two cobwebs each of length ℓ and under a tension F are contained in a glass case at temperature T. Because they are struck by air molecules they undergo random vibrations. What is the ratio of the amplitudes of these motions if cobweb *A* has twice the mass of cobweb *B*?

P143 Outdoors at night, water vapour often condenses on cobwebs, on which we can find periodical lines of very small identical water drops. Find the minimum distance between these drops.

P144* Imagine a cylindrical body that can move without friction along a straight wire parallel to its axis of symmetry, as illustrated in the figure.

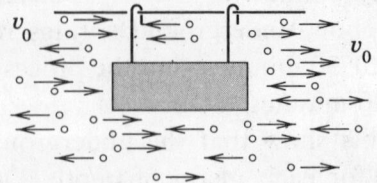

Tiny particles moving horizontally at speed v_0 bombard the body uniformly from both left and right. Collisions with the right end of the cylinder are perfectly *elastic* whilst those with the left are perfectly *inelastic*, though the particles do *not* stick to the cylinder after the collision. What is the speed of the cylinder

  (i) after a long time, and

  (ii) after a very long time?

P145 A totally black spherical space probe is very far from the solar system. As a result of heating by a nuclear energy source of strength I inside the probe, its surface temperature is T. The probe is now enclosed within a thin thermal protection shield, which is black on both sides and attached to the probe's surface by a few insulating rods. Find the new surface temperature of the probe. Determine also the surface temperature which would result from using N such shields.

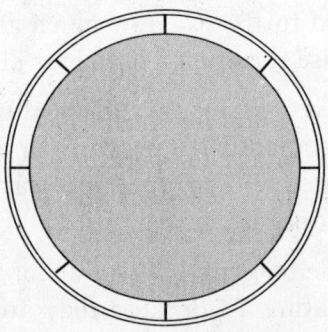

P146* Two thermally insulated containers hold identical masses of water. The water is at temperature T_1 in one of them, but at temperature T_2, $(T_2 > T_1)$, in the other. What is the maximum work that this system can do if it is used as a heat engine? Take the specific heat of water as constant over the working range.

P147 What is the change of entropy that occurs when two moles of helium and three moles of oxygen, both at s.t.p. ($T \approx 273$ K and $P \approx 1.01 \times 10^5$ Pa) and in adjacent volumes, are allowed to mix by removing the partition between them?

P148 By slowly pumping air into a 10-litre container, its pressure is increased to ten times atmospheric pressure. How much work is done during this process if the displacement of the piston in the pump is 1 litre? The walls of the container and pump are all good heat conductors and so the temperature can be taken as constant.

P149 A distant planet is at a very high electric potential compared with the Earth. A metal space ship is sent from Earth for the purpose of making a landing on the planet. Is this mission dangerous? What happens when the astronauts open the door of the space ship and step onto the surface of the planet?

P150 By what percentage does the capacitance of a spherical capacitor change when its surface is dented in such a way that its volume decreases by 3 per cent?

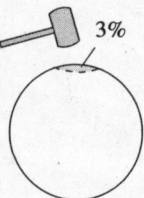

P151* A closed body, whose surface F is made of metal foil, has an electrical capacitance C with respect to an 'infinitely distant' point. The foil is now dented in such a way that the new surface F^* is entirely inside or on the original surface, as shown in the figure.

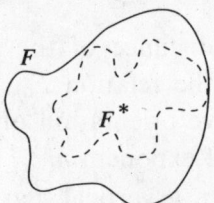

Prove that the capacitance of the deformed body is less than C.

P152   The plates of a parallel-plate capacitor have surface area A, and are initially separated by d. They are connected to a voltage V_0. What work is required to pull the plates apart to a separation of $2d$? How does the energy of the capacitor change during the process?

P153*   What is the change in length of the spiral spring shown in the figure, which has N turns, radius R, length x_0 and spring constant k, when a current I_0 is made to flow through it?

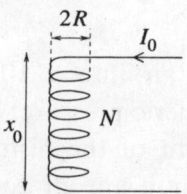

P154*   A very short magnet A of mass m is suspended horizontally by a string of length $\ell = 1$ m. Another very short magnet B is slowly brought closer to A in such a way that the axes of the magnets are always on the same horizontal level as each other. When the distance between the magnets is $d = 4$ cm, and magnet A is $s = 1$ cm away from its initial position, A spontaneously moves to attach itself to B.

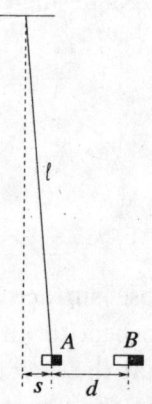

(i)   The dependence on distance of the interaction force between the magnets is given by the relation $F_{\text{magnet}}(x) = \pm K/x^n$, the sign depending on the relative orientation of the magnets. Using the given data, find the value of exponent n.

(ii)   Magnet B is placed in a vertical glass tube, which is closed at the bottom. Magnet A is then placed above it in the tube in such an orientation that the magnets repel each other. Although magnet A may tend to reverse its direction within the tube, it is constrained by the tube and cannot do so. Find the distance apart of the magnets in static equilibrium.

P155 A battery consists of N identical cells, each of e.m.f. $\mathscr{E}$. Is it true that the energy wasted when using the battery to charge a capacitor through a resistor can be reduced by charging it in N stages? That is by connecting it first across a single cell, and then across two cells, and so on, rather than across the whole battery in a single step.

P156 An 'energy-generating device' consists of a parallel-plate capacitor with nearly all the space between the plates filled with an oil of relative permittivity $\varepsilon > 1$. Calculate the stored energy in the capacitor when its plates are given charges of $\pm Q$. The oil, which cannot come into direct contact with the plates, is now poured out and replaced by air; calculate the new stored energy and show that it has increased. Explain the catch in this world-beater!

P157 An insulating sheet of relative permittivity ε_r is slowly slid between the plates of a parallel-plate capacitor, completely filling the space between the plates. What force acts on the sheet if (i) the charge, or (ii) the voltage of the capacitor is kept constant during the process?

How does the insulating sheet affect the energy of the capacitor in cases (i) and (ii)?

P158 Each element in the finite chain of resistors shown in the figure is $1\,\Omega$. A current of 1 A flows through the final element.

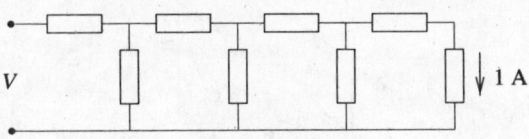

What is the potential difference V across the input terminals of the chain? What is the equivalent resistance of the chain? How does the equivalent resistance change if one or two more resistors are connected to it? Compare this result with the equivalent resistance of an 'infinite' chain.

P159 All the elements in the 'infinite' grid shown in the figure are of the same resistance R. What is the equivalent resistance between two neighbouring grid points?

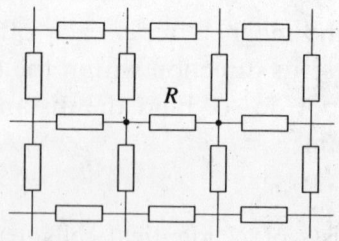

What would be the equivalent capacitance between two neighbouring grid points if all the elements in the grid were capacitors with capacitance C? What would be the equivalent inductance if the elements were inductors of inductance L?

P160* A grid in the shape of a regular polyhedron (tetrahedron, cube, dodecahedron, etc.) is made up of identical, say 1-Ω, resistors. What is the equivalent resistance between two neighbouring grid points?

P161 The previous two problems were calculations about electrical networks consisting of identical resistors (an infinite grid or a regular polyhedron). Find the equivalent resistance of a grid between two neighbouring grid points, if the resistor joining them is removed.

P162* A plane divides space into two halves. One half is filled with a homogeneous conducting medium and physicists work in the other. They mark the outline of a square of side a on the plane and let a current I_0 in and out at two of its neighbouring corners using fine electrodes. Meanwhile, they measure the p.d. V between the two other corners. This is illustrated in the figure.

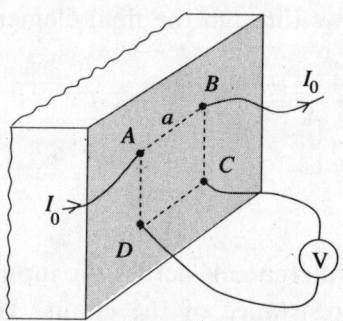

How can they calculate the resistivity of the homogeneous medium using this data?

P163* You are given a large complex electrical circuit containing a lot of resistors and other passive elements and wish to determine the resistance of a particular resistor in the circuit without unsoldering it (i.e. without taking it out of the circuit). A battery, an ammeter and a voltmeter, all of high quality, are provided. How would you carry out the measurement?

P164* All the sides of a cube are made of 1-Ω resistors. What is the equivalent resistance of the cube between the two endpoints of one of its body diagonals?

Examine one-, two- and four-dimensional 'cubes' as well. Find a general formula for the n-dimensional case.

P165 A current of 1 mA flows through a wire made of a piece of copper and a piece of iron of identical cross-sections welded end-to-end as shown in the figure.

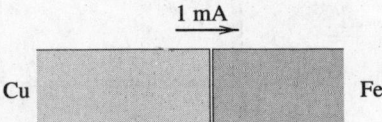

How much electric charge accumulates at the boundary between the two metals? How many elementary charges does this correspond to?

P166 The Earth's magnetic field approximates that of a dipole with a field of 6×10^{-5} T at the North Pole. Over London, the magnetic flux density is 5×10^{-5} T and the angle of dip is $66°$.

The wing span of a jumbo jet is 80 m, its length 60 m, and its depth 8 m. Estimate the potential differences that could be detected over the surface of the jet when it flies horizontally at 720 km h^{-1}:

(i) over the North Pole,
(ii) northwards over the Equator,
(iii) eastwards along the Equator,
(iv) northwest over London.

P167 A homogeneous field of magnetic induction **B** is perpendicular to a track of gauge ℓ which is inclined at an angle α to the horizontal. A frictionless conducting rod of mass m straddles the two rails of the track as shown in the figure.

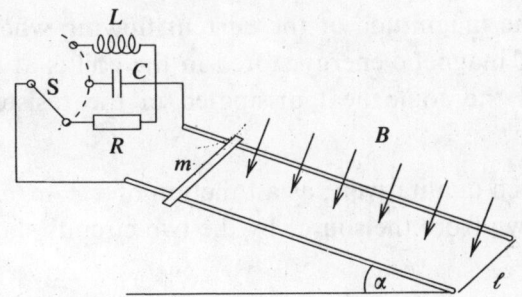

How does the rod move, after being released from rest, if the circuit formed by the rod and the track is closed by:

   (i) a resistor of resistance R,

   (ii) a capacitor of capacitance C, or

   (iii) a coil of inductance L?

P168* One end of a horizontal track of gauge ℓ and negligible resistance, is connected to a capacitor of capacitance C charged to voltage V_0. The inductance of the assembly is negligible. The system is placed in a homogeneous, vertical magnetic field of induction B, as shown in the figure.

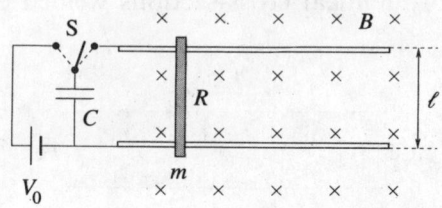

A frictionless conducting rod of mass m and resistance R is placed perpendicularly onto the track. The polarity of the capacitor is such that the rod is repelled from the capacitor when the switch is turned over.

   (i) What is the maximum velocity of the rod?

   (ii) Under what conditions is the efficiency of this 'electromagnetic gun' maximal?

P169 A resistor and an inductor in series are connected to a battery through a switch.

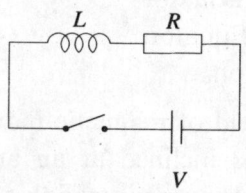

After the switch has been closed:

   (i) What is the magnitude of the current flowing when the rate of the increase of magnetic energy stored in the coil is at a maximum?

   (ii) When will the Joule heat dissipated in the resistor change at the fastest rate?

P170* (i) Sketch qualitatively, as a function of $x = \omega/\omega_0$, the magnitude of the current drawn from the source by the two circuits shown in the figure; here $\omega_0 = (LC)^{-1/2}$.

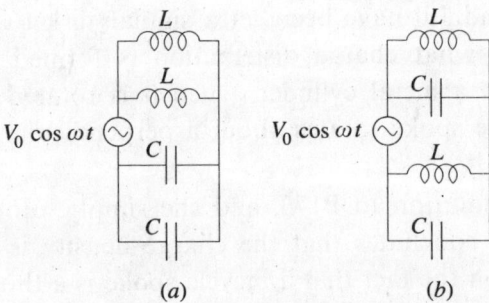

(a)   (b)

(ii) Using three or more of the components shown in figure (*a*), construct five new circuits, each of which shows current resonance (maximum current drawn from the source at some frequency), but all at different frequencies.

P171* The circuit shown in the figure – consisting of three identical lamps and two coils – is connected to a direct current source. The ohmic resistance of the coils is negligible.

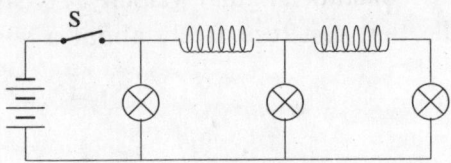

After some time, switch S is opened. What are the relative brightnesses of the three lamps immediately afterwards?

P172 The turns of a solenoid, designed to provide a given magnetic flux density along its axis, are wound to fill the space between two concentric cylinders of fixed radii. How should the diameter d of the wire used be chosen so as to minimise the heat dissipated in the windings?

P173* A solid metal cylinder rotates with angular velocity ω about its axis of symmetry. The cylinder is in a homogeneous magnetic field B parallel to its axis. What is the resultant charge distribution inside the cylinder? Is there an angular velocity for which the charge density is everywhere zero?

P174* Consider the result of the previous problem using a rotating frame of reference, fixed to the cylinder. Describe the electric and magnetic fields in this rotating (non-inertial) frame of reference.

(Assume that the angular velocity of rotation is much smaller than the cyclotron frequency, $\omega_0 = eB/m$, where e and m are the elementary charge and mass of the electron, respectively.)

P175*   Jack and Jill have been set a similar task to that in P173. They have to calculate what charge distribution is formed in a metal bicycle spoke, rather than a metal cylinder, when it is rotated in a homogeneous magnetic field. The spoke rotates about a perpendicular axis at one end of it.

Jill knows the solution to P173, and she simply adopts it. Ignoring the electron mass she concludes that the charge density is $\rho = 2\varepsilon_0 B\omega$. Jack's solution is based on the fact that a bicycle spoke is a thin metal rod; and so he considers the problem to be one-dimensional. The induced electric field is $E(r) = rB\omega$ at a distance r from the rotational axis.

Applying Gauss's law to a short section of the spoke of length Δr, Jack finds the charge density: $(\rho/\varepsilon_0)A\Delta r = \Delta E A = B\omega\Delta r \times A$, where A is the cross-sectional area of the spoke. From this equation he derives: $\rho = \varepsilon_0 B\omega$, which is only *half* of Jill's value.

Comment on these differing results.

P176   A circular metal ring of radius of $r = 0.1$ m rotates about a vertical diameter with constant angular velocity. As shown in the figure, a small magnetic needle that can turn freely about a vertical axis sits in the middle of the ring.

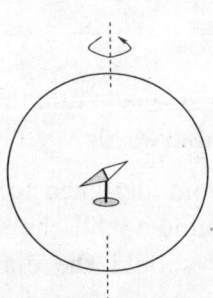

When the ring is stationary, the needle points in the direction of the horizontal component of the Earth's magnetic field. However, when it rotates at the rate of ten turns per second, the magnet deviates by an average of 2° from this position.

What is the electrical resistance R of the ring?

P177   A uniform thin wire of length $2\pi a$ and resistance r has its ends joined to form a circle. A small voltmeter of resistance R is connected by tight leads of negligible resistance to two points on the circumference of the circle at angular separation θ, as shown in the figures.

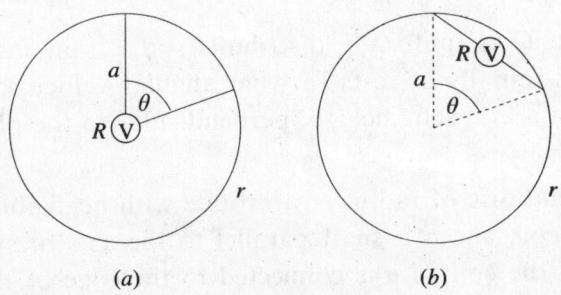

(*a*)   (*b*)

A uniform magnetic flux density perpendicular to the plane of the circle is changing at a rate $\dot{B}$. What will the reading of the voltmeter be if the voltmeter is positioned:

(*a*) at the centre of the circle, and

(*b*) on the chord joining the two points of attachment?

P178* A 'twisted' circular band (called a Moebius strip) is made from a strip of paper of length L and width d. A wire running along the edge of the strip is connected to a voltmeter, as shown in the figure.

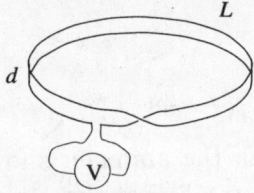

What does the voltmeter register when the strip is placed in a homogeneous magnetic field which is perpendicular to the plane of the strip and changes uniformly with time, i.e. $B(t) = kt$?

P179 A long solenoid contains another coaxial solenoid (whose radius R is half of its own). Their coils have the same number of turns per unit length and initially both carry no current. At the same instant currents start increasing linearly with time in both solenoids. At any moment the current flowing in the inner coil is twice as large as that in the outer one and their directions are the same. As a result of the increasing currents a charged particle, initially at rest between the solenoids, starts moving along a circular trajectory (*see figure*). What is the radius r of the circle?

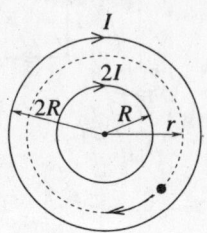

P180 Charge Q is uniformly distributed on a thin insulating ring of mass m which is initially at rest. To what angular velocity will the ring be accelerated when a magnetic field B, perpendicular to the plane of the ring, is switched on?

P181* A metal disc of radius r can rotate with negligible friction inside a long, straight coil, about a shaft parallel to the axis of symmetry of the coil. One end of the coil wire is connected to the edge of the disc and the other to the shaft. The coil has ohmic resistance R and contains n turns per unit length. It is placed so that its axis is parallel to the Earth's magnetic field vector $\mathbf{B_0}$.

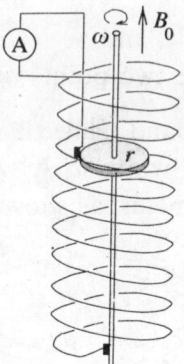

What current flows through the ammeter shown in the figure if the disc rotates with angular frequency ω? Plot the current as a function of ω for both directions of rotation.

Prove that the power needed to rotate the disc is equal to the rate of Joule heating generated by the ohmic resistance of the coil.

P182* A thin superconducting (zero resistance) ring is held above a vertical, cylindrical magnetic rod, as shown in the figure. The axis of symmetry of the ring is the same to that of the rod. The cylindrically symmetrical magnetic field around the ring can be described approximately in terms of the vertical and radial components of the magnetic field vector as $B_z = B_0(1-\alpha z)$ and $B_r = B_0 \beta r$, where B_0, α and β are constants, and z and r are the vertical and radial position coordinates, respectively.

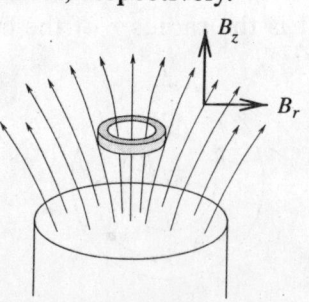

Initially, the ring has no current flowing in it. When released, it starts to move downwards with its axis still vertical. From the data below, determine how the ring moves subsequently? What current flows in the ring?

Data:

Properties of the ring:	mass	$m = 50\,\text{mg}$
	radius	$r_0 = 0.5\,\text{cm}$
	inductance	$L = 1.3 \times 10^{-8}\,\text{H}$

Initial coordinates of
the centre of the ring:

$$z = 0$$
$$r = 0$$

Magnetic field constants:

$$B_0 = 0.01\,\text{T}$$
$$\alpha = 2\,\text{m}^{-1}$$
$$\beta = 32\,\text{m}^{-1}$$

P183* A small, electrically charged bead can slide on a circular, frictionless, insulating string. A point-like electric dipole is fixed at the centre of the circle with the dipole's axis lying in the plane of the circle. Initially the bead is on the plane of symmetry of the dipole, as shown in the figure.

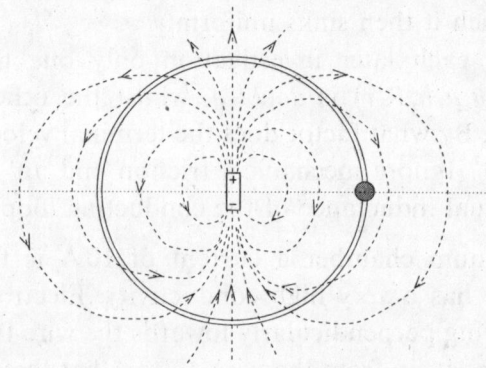

How does the bead move after it is released? Find the normal force exerted by the string on the bead. Where will the bead first stop after being released? How would the bead move in the absence of the string? Ignore the effect of gravity, assuming that the electric forces are much greater than the gravitational ones.

P184* A point-like body of mass m and charge q, initially at rest, is released in a homogeneous gravitational field. What path does the body follow if it is also acted upon by a homogeneous horizontal magnetic field?

P185* A long, thin, vertical glass tube is surrounded by a much wider coaxial glass tube of outer radius r. Wound on the wider tube there are many separate circular conducting loops, each of resistance R and spaced a distance h apart.

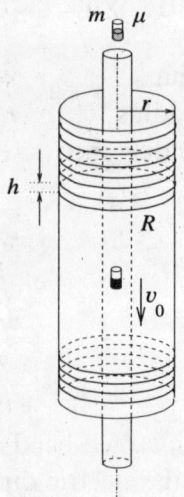

If a small magnet bar of mass m and magnetic moment μ is dropped into the thin tube, after a relatively short time it reaches a constant terminal velocity v_0, with which it then sinks uniformly.

In the course of each later investigation only one the five quantities mentioned above (m, μ, h, R, r) is *doubled*, whilst the other four remain at their original values. By what factor does the terminal velocity of the magnet change in each case? Ignore mechanical friction and air resistance, as well as the self- and mutual inductance of the conducting loops.

P186* In a vacuum chamber a current of 10 A is flowing in a long, straight wire, which has a very high conductivity. Electrons with an initial velocity v_0 start moving perpendicularly towards the wire from a point which is a radial distance r_0 away from the wire. Given that they cannot approach any closer to the wire than $r_0/2$, determine v_0. Ignore the effect of the Earth's magnetic field.

P187* The distance between the plates of an initially uncharged capacitor is d. Perpendicular to its plates, there is a magnetic field of strength B, as shown in the figure.

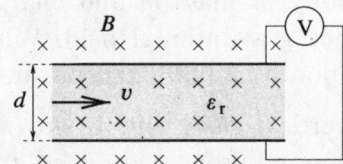

What voltage does the voltmeter connected to the plates of the capacitor register when an electrically neutral liquid of relative dielectric constant ε_r flows between the plates with velocity v?

P188 The energy released by the fission of uranium nuclei would be higher if the uranium nucleus split into three parts rather than into two. Despite this, the fission of uranium only produces two nuclei. Why is this?

P189* ^{7}Be is a radioactive element with a half-life of 53.37 days. When isotope 7 of beryllium is heated to a few thousand degrees, its half-life changes. What is the explanation for this?

P190* Part of the series of isotopes produced by the decay of thorium-232, together with the corresponding half-lives, is given below:

$$^{232}_{90}\text{Th} \xrightarrow[1.41\times10^{10}\,\text{y}]{} \,^{228}_{88}\text{Ra} \xrightarrow[5.7\,\text{y}]{} \,^{228}_{89}\text{Ac} \xrightarrow[6.13\,\text{h}]{} \,^{228}_{90}\text{Th} \xrightarrow[1.91\,\text{y}]{} \,^{224}_{88}\text{Ra} \xrightarrow[3.64\,\text{d}]{} \,^{220}_{86}\text{Rn} \xrightarrow[56\,\text{s}]{} \cdots$$

Thorium-232 and thorium-228 in equilibrium are extracted from an ore and purified by a chemical process. Sketch the form of the variation in the number of atoms of radon-220 you would expect to be present in 10^{-3} kg of this material over a (logarithmic) range from 10^{-3} to 10^3 years.

P191 Through what voltage must protons be accelerated if they are to be able to produce proton–antiproton pairs when they collide with stationary protons? The rest-mass energy of a proton is approximately 1 GeV.

P192* How does a positron move in a Faraday cage if it is 'dropped' with no initial speed? Consider the positron as a classical particle, acted on by electrical forces and the gravitational field of the Earth, as indicated in the figure.

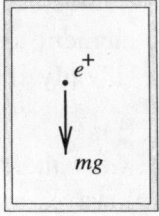

P193*   Two positrons are at opposite corners of a square of side $a = 1$ cm. The other two corners of the square are each occupied by a proton, as shown in the figure.

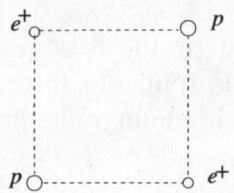

Initially the particles are held in these positions, but all four are released at the same time. What will their speeds be when they are a significant distance apart? The particles can be considered as classical point masses moving in each other's electric fields. Gravity can be ignored.

P194   In an experiment on Compton scattering, stationary electrons are bombarded by photons whose energy is equal to the rest energy of an electron. For events in which the scattered photon and the recoil electron have momenta of the same magnitude, find the angle between them. What is the speed of the recoil electron in this case?

P195   X-ray photons are scattered through an angle of 90° by electrons initially at rest. What is the change in the wavelength of the photons?

P196   Imagine a 'classical electron' as a small, spherical ball. What is its minimum radius, if its electrostatic energy is not to be greater than its total rest energy, mc^2? What is its angular velocity if its angular momentum is $h/(4\pi)$? To what 'equatorial speed' does this correspond, if the whole of the electron's rest energy is provided by the electrostatic field?

P197*   An electron is enclosed in a large rectangular box. Estimate the order of magnitude of the thickness of the layer (at the bottom of the box) which, as a result of gravitational effects, is occupied by the electron.

P198*   Classically, the Coulomb field of an atomic nucleus could confine an electron to that nucleus. However, the Heisenberg uncertainty principle prescribes such a high kinetic energy for an electron enclosed in such a small space that it would escape from the nucleus in any case. What would be the atomic number of a transuranic element able to confine an electron within its nucleus for a significant time, if only the element itself were sufficiently stable?

P199*   Show how the *size of water molecules* can be estimated using the speed of water surface (capillary) waves and the speed of sound waves in water? The speed of propagation of surface waves of wavelength 1 cm is approximately 10 000 times smaller than that of sound in water.

P200 Congratulations to the reader! You have reached the last problem in the book and the proper manner in which to congratulate you would be to drink your health *in* champagne. Unfortunately, this kind of recognition is not really practical–though we can at least make the last problem one *about* champagne.

The bubbles in champagne are familiar. They form almost exclusively at particular points in the champagne glass, and from these points they rise faster and faster. Why *do* the bubbles in champagne accelerate?

Hints

H1 Resolve the velocity vectors of the snails into suitable components. There is more than one way to do this, and they lead to different methods of arriving at the same solution. The equation of the path can be determined by expressing the velocity in polar coordinates.

H2 Calculate the maximum possible value of the coefficient of friction if the object is not to stop on the table.

H3 The solution of part (i) is trivial, since the boat is faster than the river. In part (ii), a suitably chosen vector addition can help determine the directions in which the boatman *could* go; the direction corresponding to the shortest path has still to be chosen.

H4 Although the whole of the moving part of the carpet has unit speed, its centre of mass has a lower speed. The reason for this is the increasing mass of the moving part.

H5 Draw the 'space–time' world lines of the snails. The result can also be obtained using the equivalence of different inertial frames of reference (Galilean symmetry).

H6 Compare the amounts by which the centres of mass of the two worms are raised.

H7 You can base your solution on the conservation of energy and the conditions for static equilibrium.

H8 Determine how much additional volume is submerged if the berg is depressed by a small distance x. Use the flotation condition to relate the mass M of the berg to its overall dimensions.

H9   The sum of the tensions in the suspension springs remains unaltered by parking on the pavement, and the net torque about *any* axis must be zero.

H10   The trickiest part of the solution is to appreciate how frictional forces can balance both Jean Valjean's weight and the reaction forces of the walls at the same time.

H11   The distance between the geometrical centre of the composite sphere and its centre of mass cannot be greater than a certain distance. Find this limiting distance.

H12   Describe the motion of the ball in terms of its components perpendicular and parallel to the slope.

H13   Because of the acceleration associated with its motion, the hamster exerts a force on the platform. The torque resulting from this force can balance the torque about the pivot due to the hamster's weight.

H14   Get a bicycle and try it.

H15   Use Newton's law of gravitation and express the mass of the Sun in terms of its average density.

H16   Compare the field produced by one of the stars at the position of the other, to that experienced by the Earth as a result of the Sun's gravitation.

H17   The space probes should be launched from the Equator and directed eastward.

H18   Examine the energies involved.

H19   The centre of mass of the system, i.e. honey plus steel ball, moves steadily downwards. The total momentum of the system can be calculated using the speed of the centre of mass, and the momentum of the honey obtained by then subtracting that of the ball.

H20   If the wall of the container is at a temperature different from that of the gas, then collisions of the gas molecules with the wall either take energy from the wall or give energy to it.

H21   The difference in temperature between the two spheres arises because their centres of mass are displaced in opposite directions.

H22   No student reading this book should need a hint.

H23   The situation is possible – and using only two resistors!

H24   Examine how the combined centre of gravity of the bucket and water changes as the water leaks away.

H25   Under what conditions does adding a little more water inevitably raise the overall centre of gravity?

H26   Our assertion is that water does not flow into the bowl. In order to prove this, the effects of both the force and the torque exerted by the chain have to be considered.

H27   The only unusual part of the solution is the calculation of the buoyancy force.

H28   The average density of the bubble has to be the same as that of air, since the bubble floats.

H29   At the end of the capillary tube, the pressure of curvature balances the difference between the pressure inside the liquid and atmospheric pressure.

H30   The whole system (the current distribution and the electric field) is spherically symmetrical, and therefore the magnetic field also has to be. Consider which spherically symmetrical magnetic fields are consistent with the (experimentally observed) non-existence of magnetic monopoles.

H31   Make use of the symmetry of the charge distribution.

H32   It is not sufficient to simply compare the gravitational accelerations, since it isn't clear that the high-jumper would be able to take off with the same initial speed. The movement of the high-jumper's centre of mass during the jump has to analysed.

H33   The time intervals of the motions and the lengths of the paths do not have to be found exactly; only the inequalities relating them need to be determined.

H34   Resolving the tension in the string into radial and tangential components, the direction of the string can be calculated using the dynamical conditions for uniform circular motion.

H35   Prove that, at any instant, bodies which started at the same time, from the same point and slid down frictionless wires in different directions, all lie on the surface of a common (imaginary) sphere.

H36   The problem can be solved in an elementary way using a rotating frame of reference fixed to the minute hand.

H37   The stone moves away from the thrower until the component of its velocity parallel to its position vector has decreased to zero. If this never occurs, the condition imposed in the problem has been met.

H38 It is *false* to assume that the trajectory of the grasshopper (with the minimum take-off speed) just touches the trunk at its topmost point.

H39 They clearly cannot jump directly towards each other without mishap, so consider jumping in some other direction whilst preserving the symmetry of the situation. Note that the mass of the hair is given!

H40 The shape of the common surface of the water jets, their envelope, has to be determined. They start from the same place, have the same initial speed, and follow parabolic paths. Examine the condition for determining whether any water jet passes through a given point in space.

H41 Show that

$$\text{Range} = \frac{v^2}{g}\left[\sin 2\theta + \frac{EQ}{mg}(1 - \cos 2\theta)\right],$$

and maximise with respect to θ.

H42 The normal reactions exerted on the rod by my fingers are not equal in general. Thus the maximum static frictional force is smaller on one side than on the other, and sliding will occur there first. However, because of the increasing normal reaction at the finger where sliding is taking place, the kinetic frictional force there increases, and the moment it becomes larger than the static friction on the other side, slipping will stop at the first finger and start at the second. During the process there is alternating slipping and sticking at both fingers. The work done can be calculated from the length of each stage.

H43 The process should be started from the top! The correct strategy is to slide the topmost brick as far as possible and then do the same thing with the two uppermost, considered as a unit, and so on downwards.

H44 Use the balances of forces and torques acting on the plate to find connections between the frictional forces involved. A graphical method of treating the linear equations so derived is recommended.

H45 The process has to be considered as a series of consecutive collisions.

H46 Show that in the first collision the fraction of the initial kinetic energy transferred to the middle ball is $4\mu M/(\mu + M)^2$.

H47 During the collision, the momentum, energy and angular momentum of the system are all conserved.

H48 Decompose the motion into that *of* the centre of mass and that *in* the centre of mass system. Show, by considering the conservation of energy and angular momentum, that when the rope first tightens the centre of mass velocities of both particles are unchanged in magnitude but turned through $\pi/2$, with the result that they then travel parallel to the x-axis.

H49 The usual reasoning which assumes that one-third of the basin is filled in 1 minute, and one-half of the basin empties in the same time, (and hence that $\frac{1}{2} - \frac{1}{3} = \frac{1}{6}$ of the basin becomes empty in 1 minute), is false. Water flows into the basin uniformly from the tap, but (according to Torricelli's law of efflux) it flows out more quickly when the water level in the basin is higher.

H50 Show that the free surface is part of the paraboloid of revolution $z = \omega^2 r^2/2g$, where z is measured from the lowest point of the free surface and r is the radial distance from the central axis. Consider the volume of the air above the liquid but still inside the vessel.

H51 In the context of mechanics, the car is *not* a closed system; it is in contact with its surroundings, in this case, the Earth.

H52 The focal length can be obtained using the relationship between the lens formula and the magnification. The ratio of the brightness values depends not only on the size of the images, but also on the amount of light reaching the lens.

H53 The apparent magnitude of the virtual image is not determined by the size of the image itself, but by the angle it subtends at the eye.

H54 Obtain $n_g \sin(\theta + \phi) \geq n_w$, where ϕ is the angle in the glass between the ray and the normal to the surface at the point where it enters the prism.

H55 No patch of light can be seen either right next to the quarter-cylinder, or very far from it. The closer light patch is excluded by total internal reflection. The distance of the furthest part of the light patch can be determined by considering the part of the quarter-cylinder close to the table as a plano-convex lens.

H56 Suppose that the sunlight falling on the Moon is diffusely reflected with the given coefficient. Calculate how much of it reaches unit area of the Earth.

H57 The most comfortable walking rate can be related to the period of the human leg swinging freely like a pendulum. Running can be considered as a forced oscillation, with its period dependent on the moment of inertia of the leg and the torque applied by the muscles.

H58 By choosing its length suitably, a simple pendulum can be made to have the same angular velocity, in any position, as that of the rod pendulum given in the problem. Compare the periods of swing of this pendulum and the actual simple pendulum of the problem. The ratio of the periods of two simple pendulums of different lengths displaced through the same angle can be deduced using dimensional analysis.

H59 Identify the physical quantities on which the power necessary for hovering depends.

H60 Use conservation of energy to determine the rod's angular velocity ω when its inclination is θ, and relate the components of the reaction between the table and the rod to the accelerations they produce. In case (i), the smooth horizontal and vertical walls of the groove can exert only vertical and horizontal forces on the end of the rod, respectively. In case (ii), the edge of the table is a very small quarter-circle, so the normal force is always directed along the rod's axis.

H61 When the coefficient of friction is small, the point of the pencil moves 'backward'. If the coefficient of friction is larger than a certain critical value (which can be shown to be about 0.37), the pencil moves 'forward'. Using the fact that kinetic friction decreases the mechanical energies, it can be shown that the point of the pencil never loses contact with the table.

H62 Use the ideal gas equation to express the conservation of air mass. Also note that, after a sufficiently long time, the temperature of the system will not have changed.

H63 Because of the surface tension (pressure of curvature) of the water, the pressure inside the trapped water is lower than atmospheric pressure.

H64 Calculate the velocity of the points of the thread at any given moment.

H65 Consider the ('elastic') frame of reference fixed to the thread.

H66 Find a simple – mathematically easy to treat – trajectory, in which the ball reaches as high a speed as possible, and the time so gained compensates for the longer path involved.

H67 From the figure you can determine the angle *not* given in the text.

H68 The centre of mass (CM) of the compasses is directly below the attachment point. If the angle between the arms were changed, the horizontal position of the CM would have to remain the same, although the positions of the CM of the individual arms would change. Use this argument to find the solution to the problem with a minimum of actual calculation.

H69   Express the condition for equilibrium using vectors.

H70   Note that the system as a whole has no external horizontal forces acting upon it. As well as keeping the centre of mass of the system fixed, you will need to use the law of conservation of linear momentum. You should be able to show that the tanker will initially move forward, but later reverse its direction of motion.

It may seem surprising that the tanker alters its direction of motion, and it might help to first consider the following, rather than the original problem:

> A poor student and a zealous ticket collector, both of mass m, are in a stationary, frictionless railway carriage of mass M. When the collector realises that the student has no ticket, the student runs to the end of the carriage with the collector, who moves with speed v relative to the carriage, in pursuit. The student stops at the end of the carriage and jumps out. Find the velocity of the carriage when the ticket collector reaches the open door, stops there and watches the student making his escape.

H71   Consider the motion in the frame of reference of the common centre of mass of the beads.

H72   In the case of the inelastic collisions, after a sufficiently long time, a growing mass of the coalescing beads reaches a constant velocity. Apply Newton's law of motion to this cluster. For elastic collisions, first examine what would happen if the external force acted only until the first collision had occurred.

H73   Consider what is happening to the centre of gravity of the table plus jug plus beer.

H74   The viscosity of the water can be taken to be small and the change in potential energy of the liquid should be neglected compared with the kinetic energy. Note that the gutter cannot change the horizontal momentum of the jet of water and Bernoulli's equation is applicable (several times!).

H75   Obtain expressions for the changes, over a very short time interval Δt, in the potential and kinetic energies of the initially stationary liquid, as it starts to move with (initial) acceleration a.

H76   Experience indicates that the rate at which the sand runs through the constriction does not depend upon the amount of sand in the upper part of the egg-timer. The explanation for this is that, due to the friction between the grains of sand, the average speed of the emerging sand depends only on its nearby environment, primarily on the diameter of the hole, and not on effects originating from remote parts. (This is not true for liquids, where pressure effects are transmitted through large distances; *see* P49.) Thus the time that the sand takes to run through the hole has to be proportional to the cube of the initial height H of the sand. Find the other quantities on which this time may depend and then apply the method of dimensional analysis.

H77 For small displacements the net force exerted on the bob is $F(x) \approx -kx^3/\ell_0^2$, where k is the spring constant. Using dimensional analysis one can deduce the dependence of the period on the spring constant, the mass of the bob and the amplitude of its motion.

H78 In the given circumstances, both the horizontal and the vertical motion of the body can be approximated by harmonic oscillations.

H79 Describe the motion in the (decelerating) frame of reference of the train.

H80 Examine the motion in the frame of reference fixed to the wedge.

H81 Under what conditions would a long, thin thread move uniformly above the Equator in a synchronous orbit, i.e. with the same angular velocity as the Earth?

H82 The normal component of the acceleration of the car is $a_n = v^2/\rho$, where v is the speed of the car and ρ is the radius of curvature of the bridge. The latter can be deduced by considering the motion of a projectile; it follows a trajectory which has the same shape as the surface of the bridge.

H83 Determining the radius of curvature of the track is the core of the solution (*see* H82).

H84 In any time interval, the water carries the boat downstream by the same amount as the remaining distance to the mark has been reduced.

H85 Calculate by how much the speed of the pushed child and its velocity component down the slope change in unit time. Find a relationship between the rates of change of these two quantities.

H86 Compare the rate of decrease of the distance between the smugglers' ship and the coastguard's cutter, to the speed at which the latter moves away from the shore.

H87 Because of the symmetry of the problem, the bodies are always at the corners of an ever-decreasing regular n-gon, and each of them moves as if only the gravitational attraction of a centrally placed single body (of a suitably chosen mass M_n) acted on it. The time taken for the system to collapse into the centre can be calculated using Kepler's third law.

H88 The solution needs the application of *all three* of Kepler's laws of planetary motion.

H89 Recall the strengths and directions of the field associated with a magnetic dipole; $B_\parallel = 2\kappa\mu/L^3$ on its polar axis (*A* position of Gauss) and $B_\perp = \kappa\mu/L^3$ on its equator (*B* position of Gauss), where κ has been written for $\mu_0/4\pi$.

H90 The force acting on the charge can be found using the so-called method of image charges. The force – analogous to gravitational attraction – is inversely proportional to the square of the distance. Therefore, the body's behaviour is similar to the motion described by Kepler's laws for a degenerate elliptical orbit.

H91 Use the method of image charges to find the value of the electric field and the induced surface charge density in the region below the ball. The 'negative pressure' due to the electrostatic forces acting on the surface of the brine below the ball is balanced by the hydrostatic pressure of the water 'hump'.

H92 Apply the method of spherical image charges. The basis of this method is that the electric field produced by two point charges, of opposite signs and different absolute values, has a sphere as its zero potential surface.

H93 You need to use both the laboratory and the centre of mass reference frames.

H94 If the sequence of events were re-played in slow motion, it would be seen that immediately after the collision, the first ball stops and rotates in a fixed place, whilst the second ball moves on but without rotation. Thus, in the overall collision, the first ball transfers linear but not angular momentum to the second ball.

After the collision, friction moves the first ball forward, but slows its rotation. On the other hand, friction slows the translational motion of the second ball, whilst increasing its rotation. Thereafter, the angular momentum of each of the balls about its point of contact with the table remains constant.

H95 Consider energy conservation, but don't forget to include dissipation as heat.

H96 Examine the angular momentum of the ball about a point on the table which lies on the ball's path, but is otherwise arbitrary.

H97 Convince yourself that it is only the east–west component of the traffic momentum that matters.

H98 Dynamical equations of motion and connections between translational and angular accelerations are central to the solution.

H99 From the point of view of energy and momentum, the system is not closed and, therefore, these quantities are not conserved. Conservation of angular momentum explains the strange phenomenon described in the problem.

H100 Consider the forces acting upon a short length of the ring which subtends an angle $\Delta\theta$ at the axis of rotation.

H101 Determine the difference, ΔF, in the force stretching the thread around the surface of the cylinder at two points on the cylinder's surface whose azimuthal separation is $\Delta\alpha$. This change in the force is proportional to the force acting normally on the cylinder, and this in turn is proportional to F. Consider an equivalent phenomenon, in which the rate of change of some quantity is proportional to the quantity itself, (e.g. radioactive decay, capacitor discharge, etc.). Using the analogy, relations applying to the friction of the thread can be obtained.

H102 Jenny suggests that Charlie considers a homogeneous ring rotating with constant angular velocity about an axis perpendicular to its plane and passing through its centre. He should determine which forces act on the ring, and consider how Newton's second law is satisfied for the centre of mass of a piece cut from the ring.

H103 The gravitational force both accelerates the hanging part of the chain and impulsively sets into motion the next link. This means that the changing mass of the moving chain has to be taken into account.

H104 Gravitation can be ignored in the frame of reference moving with the centre of mass of the chain; in this frame the chain will be weightless, but not massless. Examine in which directions forces act on a small piece of the chain, which has radius of curvature R and moves at a uniform speed v, and consequently determine how the shape of the chain is deformed (*see also* P100, P101 and P102).

We can reveal that Frank's guess is right – the chain keeps its original shape.

H105 Calculate the tension in the chain when it leaves the pulley. Use the principle of conservation of energy (*see also* P104).

H106 Examine the motion of the loop in the frame of reference moving at the same speed c as the centre of the loop. In this frame the pieces of the loop travel with uniform circular motion. The conditions governing the dynamics of circular motion yield an equation for c.

H107   Examine the change in the horizontal momentum of the sand falling onto the belt in unit time. Consideration of the energies involved is also useful.

H108   Apply the law of conservation of energy; then find the force by using the change in the momentum calculated from the speed of the roll as a function of its position.

H109   The gravitational field of a thin spherical shell of uniform mass distribution is zero inside the shell. Outside the shell, it is the same as if the total mass of the shell were concentrated at its centre.

H110 – The gravitational field inside a homogeneous sphere is directly proportional to the radius of the sphere (*see* P109). The gravitational field of the hollowed-out sphere can be found by superimposing the fields of a homogeneous sphere and a smaller sphere of 'negative mass density'.

H111   Divide the hemisphere into equally thick hemispherical shells. Prove that these shells each produce the same gravitational field at the point in question.

H112   Calculate the force a 'mythical giant' would have to exert to pull the two halves of the asteroid (already cut in two) apart by 1 m.

H113   The electric field exerts a force whose magnitude is proportional to the surface area exposed by the cut and is in a direction perpendicular to that surface. Note that this force is similar to that caused by liquid or gas pressure.

H114   At first sight several parameters seem to be missing. Don't worry about it! Find the equilibrium condition and calculate the electrostatic energy of the system in that situation.

H115   Find the maximum amount of hydrogen that the container could have contained initially without bursting. The material from which the container is made may be chosen freely, but only from real materials.

H116   The laws of gravitational and electrostatic fields are very similar. Make use of this similarity and apply Gauss's law.

H117   Examine the electric field due to a rod element which subtends an angle $\Delta\alpha$ at the point C.

H118   Consider two very long rods joined end-to-end.

H119 You can distinguish between the correct and false formulae by considering the case in which θ approaches π. Apply the well-known expression for the magnetic field of a long straight current-carrying wire to find the proportionality factors.

H120 Imagine that another identical coil is joined symmetrically to the original solenoid at point P, and that the same current is also allowed to flow in this second coil. Apply the principle of superposition.

H121 It is easy to find the force if one imagines changing the positive charges on one of the plates into negative charges of the same magnitude. On the other hand the electric field line structure of the positive–positive plates is very different from that of the positive–negative arrangement.

H122 Remember that the total charge on an isolated plate cannot change.

H123 The total charge induced on each plate would not change if the point charge Q were considered to be spread uniformly over a plane a distance x from the lower plate.

H124 The total electric field outside the plates must be exactly zero. What are the consequences of this well-known fact for the charge distribution?

H125 Imagine that the medium is sliced into thin layers perpendicular to the y-direction. The individual layers can be considered as plane-parallel plates with different refractive indices, and the relation between the refractive index of a layer and the angle of incidence of the light ray can be determined.

H126 Using simple geometry, you can measure the useful surface area of a CD. To obtain the required result divide this area by 650 M and also by 8, because 1 byte = 8 bits. You can treat a CD as a reflection grating and measure its diffraction pattern using a laser beam of known wavelength.

H127 Determine $n\lambda$ for the composite line and consider possible values of n, the order of the diffraction spectrum.

H128 In case (i), the optical path difference consists of two parts; one originates in front of the grating and the other behind it.

In case (ii), instead of considering an optical grating, investigate the diffraction pattern from a single slit, which is tilted 'forward' through an angle ϕ.

H129 Draw diagrams showing liquid levels and pressures in the space between the objects and on either side of them.

H130 Find the horizontal forces acting on the meniscus.

H131 During evaporation, the surface area of the drop shrinks, and its surface energy decreases. Compare this energy decrease with the energy needed for evaporation.

H132 The equilibrium saturated vapour pressure is slightly higher near the surface of a smaller drop than near the surface of a larger one. The vapour pressure is uniform at the bottom of the container and its value is therefore higher than the equilibrium value for large drops, but lower than that for small drops. Consequently, vapour evaporates from the smaller drops, making them smaller, and condenses onto the large ones, making them larger.

The relationship between the equilibrium pressure of saturated vapour and the curvature of the drop can be deduced by considering the pressure balance in a vessel containing a capillary tube hanging into some of the liquid.

H133 The increase in internal energy of the air enclosed in the container is equal to the decrease in potential energy of the load hung from the piston.

H134 If a mountain is very high then its base melts because of high pressure. Compare the energy needed to melt the bottom layer of a mountain with the gravitational energy that would be released if the mountain then sank.

H135 If the state of the enclosed air is plotted on a p–V diagram, a straight line is obtained. The hidden elegance of the problem is revealed when the implications of the straight line's being tangential to an isothermal or adiabatic curve at certain points is realised.

H136 Your explanation should be based on the interaction between the molten magma and the ice.

H137 The hydrostatic pressure of the water in the flue increases the pressure of the water in the cavity, and so it boils at a temperature higher than the usual $100\,°C$. The relationship between the pressure and temperature of the saturated water vapour can be obtained from tables or by using the approximate law

$$p = A\mathrm{e}^{-L_\mathrm{m}/(RT)}.$$

Here p is the pressure of saturated water vapour at its boiling point T, L_m is the molar heat of vaporisation of water, R is the gas constant and A is a constant with the dimensions of pressure. When the geyser erupts, the superheated water in the cavity reaches equilibrium again by boiling until it cools down to $100\,°C$.

H138 Consider the heat balance at the base of the layer when the layer thickness is x.

H139 Establish that the time taken varies as the square of the linear dimensions for similarly shaped bodies.

H140 The 'trap' hidden in this problem relates to the heat of vaporisation. The heat of vaporisation of water at $100\,°C$ and a pressure of 1 atm (the standard value of 2256 kJ kg^{-1} found in tables) takes into account not only the higher internal energy of the vapour but also the work done by expansion against atmospheric pressure.

H141 A liquid starts boiling when its saturated vapour pressure reaches or surpasses the pressure of the gas above the liquid.

H142 Consider how T could be incorporated in a formula for the amplitude.

H143 Compare the surface energy of a long cylinder of water (assuming that the cobweb is uniformly covered with water) and the surface energy of the periodic water drops.

H144 The cylinder keeps accelerating until the net momentum received per unit time, due to the particles colliding with it from both the left and the right, becomes zero. However, after a very long time, the cylinder stops moving, in agreement with the second law of thermodynamics.

H145 Take into account both the emission and the absorption of heat by the space probe and the consecutive inner and outer surfaces of its protecting shields.

H146 The entropy of the system cannot decrease during the process.

H147 Consider entropy from the point of view of the number of microstates available.

H148 Calculate the change in entropy of the air that is pumped into the container.

H149 The electric field strength is zero inside the space ship, just as it is inside a Faraday cage. Examine whether the electric potential of the space ship changes during the journey.

H150 Examine the change in energy of the spherical capacitor when it carries a set charge.

H151 Compare the energies of the electrostatic fields of the dented and undented foils.

H152   The opposite charges on the capacitor plates attract each other, and therefore work has to be done to pull the plates apart. The capacitance of the capacitor decreases, and as, at a given voltage V_0, the electrostatic energy of a capacitor is proportional to its capacitance C, the energy of the capacitor *decreases*! The solution to this paradox is that a capacitor connected to a battery cannot be considered as a closed system.

H153   Because the current flows in the same direction in each turn, the spring contracts. The force of contraction caused by the current can be found by considering a superconducting spiral spring (at a very low temperature in practice). A current can flow in such a superconducting coil even if its ends are short-circuited. Examine the dependence of the energy of this closed system on its length.

H154   Find the net force (the sum of the magnetic force, the weight and the tension in the string) exerted on magnet A as a function $F(x)$ of the distance x apart of the magnets. Use $F(x)$ to determine the conditions for equilibrium and stability.

H155   Calculate the total work done by the battery.

H156   Remember that the charge is unchanged when the oil is removed.

H157   The energy (per unit volume) of the electrostatic field is proportional to the square of the electric field strength and to the dielectric constant of the medium: $W_{el} = \frac{1}{2}\varepsilon_0\varepsilon_r E^2$. The dielectric between the plates of the capacitor decreases the electric field (as a result of its polarisation), and therefore the energy of the system decreases as well. The force acting on the dielectric can be calculated from this change in energy (using the work theorem).

H158   Apply Kirchhoff's laws, starting from the final element of the chain. Look for a relationship between the currents flowing through the consecutive resistors and the terms in the Fibonacci series.

H159   Consider two different cases. In the first case, a current I flows into a grid point. In the second case, a current I flows out of the neighbouring grid point. In both cases make use of the symmetry of the system, and then superimpose the two current and voltage distributions.

H160   Apply the method of superposition as in the previous problem. Be careful, since with a finite grid the current has to flow out of the circuit somewhere in order to conserve charge. Solve this difficulty without spoiling the symmetry of the problem.

H161 The key phrase of the solution is 'in parallel'.

H162 As in the previous three problems, superposition is a great help.

H163 The battery is to be connected to the terminals of the resistor through the ammeter. You need to ensure somehow that *all* the current measured by the ammeter flows through the particular resistor, and not partly via other electrical elements.

H164 Establish sets of equipotential points on the cube, when a current I is flowing in at one end of the diagonal and out again at the other end. The circuit can then be simplified by notionally connecting together all points at the same potential.

H165 Apply Gauss's law.

H166 Use Fleming's right-hand rule.

H167 The rod starts to accelerate down the slope under gravity. Electromagnetic induction causes a current to flow in the rod, which in turn brakes its motion according to Lenz's law. The equation of motion of the rod (written in terms of the current) is the same in all three cases. The different behaviours are due to the different relationships between the current flowing in the rod and the induced electromotive force.

H168 The change in velocity of the rod is directly proportional to the change in charge of the capacitor. The rod accelerates until the induced e.m.f. balances the remaining voltage across the capacitor.

H169 You can answer question (i) without solving the differential equation for the circuit. Express the rate of increase of magnetic energy as a function of the current.

For (ii), note that the time dependence of the current for this circuit is well known and that the Joule heat is directly proportional to the square of the current. If you sketch a graph of the square of the current as a function of time, you can find qualitatively the time at which the rate of change of dissipation in the resistor is the fastest. Using the result from (i), the quantitative answer can be found without the need to use calculus.

H170 Consider first a circuit containing only one inductor and one capacitor connected in series, and show that it has current resonance at a particular frequency.

H171 According to the law of induction, the current flowing in a coil cannot change *suddenly*.

H172   Show that n, the number of turns per unit length of the solenoid, is proportional to d^{-2}, and consider the resistance of one turn.

H173   Examine the forces keeping the electrons in the metal in circular orbits. If the electric field strength is known, Gauss's law can be used to determine the corresponding charge distribution.

H174   The electric field can be defined by the force acting on a unit charge, and the magnetic field can be interpreted with the help of the Lorentz force exerted on a moving charge.

H175   Jill's result is correct, and Jack's answer is wrong. The crucial point is that the electric field lines within the rotating spoke are not parallel.

H176   The magnetic field of the Earth induces a current in the rotating ring, which changes the average magnetic field at the centre of the ring. As a result the magnetic needle moves.

H177   Let the current through the voltmeter be I and that through the major arc of the ring be i. Then, using consistent conventions for current directions and circuit traversal, apply Kirchhoff's laws to two different closed circuits.

H178   A Moebius strip is a surface that has no associated direction and the law of induction must be applied only with great caution. Imagine the wire marking the edge of the band to be laid out on a plane in such a way that it does not cross itself. Determine the area of the plane enclosed by the wire and find its equivalent for the Moebius strip.

H179   Consider not only the magnetic field and magnetic forces acting on the charged particle, but also the effects of the induced electric field.

H180   As a result of the electromagnetic induction an electric field is established in the charged ring, and its tangential component causes the ring to experience a torque. One can show that the final angular velocity of the ring depends only on the final field, and not on the way it is turned on.

H181   An electric field is induced in the rotating disc and this induces a current in the coil. The total magnetic field is that due the Earth, increased or decreased by that due to the coil, according to the direction of the rotation.

H182   The total magnetic flux through the superconducting ring (consisting of that due to the external field and its own flux) must not change during the motion. The flux of the external field changes during the motion, but the change is balanced by the magnetic flux due to the current induced in the ring. If the current is known, the Lorentz force can be calculated, and the net force acting on the ring can be found as a function of its position. This resulting equation of motion is similar to a well-known mechanics equation.

H183 The electrostatic field of a dipole can be calculated from its potential $\Phi = K \, (\cos\theta/r^2)$, where K is a constant proportional to the strength of the dipole, r is the distance from the dipole and θ is the polar angle measured from the dipole's axis. Calculate first the normal force exerted by the string on the bead.

H184 Moving with velocity v_0 perpendicular to a magnetic field of magnitude B, is equivalent to being in an electric field of magnitude $v_0 B$. If v_0 is suitably chosen, this electric field can be made to cancel the gravitational field acting on the particle.

H185 The changing magnetic field induces eddy currents in the loops, which brake the fall of the magnet. The terminal speed clearly depends on the resistance of the conductors. The dependence on other parameters can be found by applying dimensional analysis. Don't forget that formulae involving magnetism usually contain the vacuum permeability μ_0, which has non-trivial dimensions.

H186 Although it is possible to solve this problem in a reference frame fixed to the vacuum chamber, the solution is rather complex. It is much easier to handle the problem using a different frame of reference that moves with velocity v_0 parallel to the wire. In this frame of reference the motion of electrons is subject to both an electric field and a magnetic field. On the other hand, in this frame, when the electron is closest to the wire its velocity is zero and the work–energy theorem can be used to solve the problem.

H187 Describe the phenomenon in the frame of reference of the liquid. Consider first the transformation of the electric and magnetic fields when they are viewed in two different frames of reference, one moving at speed v_0 relative to the other. Take v_0 as being much less than the velocity of light and ignore relativistic effects.

H188 Consider the initial (activation) energy required for three fission products rather than two.

H189 ^{7}Be decays via electron capture.

H190 In equilibrium, abundances are proportional to half-lives. Use this to show that the equilibrium value for the number of radon-220 atoms is 3.3×10^5. Then consider how this element can be produced from the purified sample.

H191   Use the relativistic formulae for energy and momentum conservation.

H192   Examine the motion of the electrons in the wall of the Faraday cage.

H193   The mass of the proton is much (nearly 2000 times) larger than that of the positron. For this reason, the positrons move with a much larger acceleration than the protons and there will be a period in the motion when the positrons have already moved far from the square, whilst the protons have hardly moved at all.

H194   Apply the (relativistic) conservation laws of energy and momentum.

H195   In the course of this process (Compton scattering), the total energy and momentum of the colliding particles (photons + electrons) remains unchanged. It is convenient tò take the rest energy of the electron $E_0 = m_e c^2 \approx 510$ keV as the unit of energy for the calculations.

H196   Consider the electron as a spherical capacitor with radius r and a uniform surface charge distribution. The moment of inertia of a sphere of mass m and radius r is $I = Kmr^2$, where K is a dimensionless constant depending on the mass distribution. For example, for a homogeneous sphere $K = \frac{2}{5}$.

H197   Apply the Heisenberg uncertainty principle and consider the total energy of the electron.

H198   When an electron is enclosed in a sphere of radius r, the uncertainty principle prescribes a minimum momentum for it of $p \approx \hbar/r$. Using approximate relativistic formulae, calculate the total energy (the sum of the electrostatic and kinetic energies) of the electron as a function of the radius r and find the minimum of this function.

H199   For small wavelengths, the speed of propagation of surface water waves is determined by the surface tension. Examine the dependence of the speed of these (capillary) waves on their wavelength and consider whether this implies a lower limit for the wavelength.

H200   Take a bottle of champagne and try it. If you do not drink too little (or too much!) of it, you will almost certainly spot the reason for the acceleration.

Solutions

· **S1** Resolve the velocity of snail 2 into a component pointing towards snail 1 and a component perpendicular to this (*see* Fig. S1.1). These two snails approach each other at a relative speed of $v + \frac{1}{2}v = \frac{3}{2}v = 7.5$ cm min^{-1}, and therefore they meet after a time given by 60 cm/7.5 cm min^{-1} = 8 min. In fact, they must all meet after this time and, as they actually travel at a speed of 5 cm min^{-1}, they each cover a distance of 40 cm before doing so.

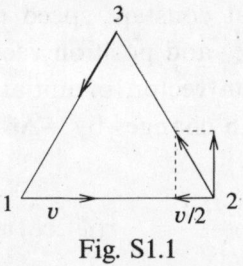

Fig. S1.1

The same result can be obtained if the velocity vector of one of the snails is resolved as shown in Fig. S1.2 into a component pointing towards the centre of the triangle formed by the snails, and a component perpendicular to this. This shows that the snails approach the centre of the triangle (it is obvious that this is where they meet) at constant speed $(\sqrt{3}/2)v = (5\sqrt{3}/2)$ cm min^{-1}, whilst travelling around this point at a tangential speed of $\frac{1}{2}v$.

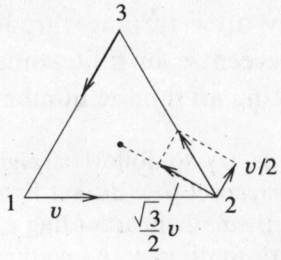

Fig. S1.2

It is easy to show that the snails are initially at a distance of $60\,(\sqrt{3}/3)$ cm from the centre of the triangle, and that therefore they meet in

$$\frac{60\,(\sqrt{3}/3)\ \text{cm}}{5\,(\sqrt{3}/2)\ \text{cm min}^{-1}} = 8 \text{ min.}$$

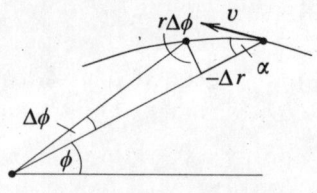

Fig. S1.3

Because of the geometrical symmetry of the situation, each snail always moves so that its direction of motion makes an angle of $\pi/6$ with the line joining its current position to the centre of the triangle. However, it is worth generalising the problem of calculating the trajectory. Consider the motion of a body moving at constant speed v around a fixed point with the angle between the velocity and position vectors equal to a fixed value α, $(0 < \alpha < 90°)$. If the position vector, of initial length r_0, moves through a small angle $\Delta\phi$ and its length changes by $-\Delta r$ (*see* Fig. S1.3), then, since α remains constant,

$$\frac{\Delta r(\phi)}{\Delta\phi} = -r(\phi)\cot\alpha.$$

This equation is very similar to the radioactive decay equation, $dm(t)/dt = -m(t)\,\lambda$, the known solution of which is $m(t) = m_0\,e^{-\lambda t}$. Using this analogy, the equation of a snail's path (in polar coordinates) is

$$r(\phi) = r_0\,e^{-\phi\cot\alpha}.$$

This is the equation of the so-called logarithmic spiral and implies that the radius r tends to zero only after turning through an infinite angle, i.e. a point-like body reaches the centre in finite time and by covering a finite distance, but only after making an infinite number of turns about the centre.

> *Note.* Nocturnal insects try to follow straight flight paths by keeping a constant bearing with respect to a *distant* light source (e.g. the Moon). If a nearby lamp misleads them, then, according to the solution just found, they will follow a spiral path to disaster. As neither the insects nor the lamp are point-like, sooner or later the insects hit the lamp.

S2 The average speed of the object is $\frac{1}{2}$ m s^{-1}. Since it decelerates uniformly, $v_{\text{average}} = \frac{1}{2}(v_{\text{initial}} + v_{\text{final}})$, and thus its initial speed cannot be greater than 1 m s^{-1}, (since $v_{\text{final}} \geq 0$). It follows that the speed of the body decreases by a maximum of 1 m s^{-1} in 2 s. Thus the absolute value of its acceleration $|a|$ is at most 0.5 m s^{-2}, i.e. $\frac{1}{20}$ times the gravitational acceleration. Therefore, the coefficient of kinetic friction between the object and the table surface cannot be greater than $\frac{1}{20}$. This is much smaller than the coefficients of friction between ordinary materials and therefore it is very likely that the object does not slide, but that all or part of it *rolls*.

S3 (i) The shortest path is one perpendicular to the bank and the boat goes in this direction if the boatman rows in the direction shown in Fig. S3.1.

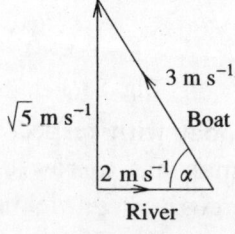

Fig. S3.1

The resultant speed of the boat (in the direction perpendicular to the bank) is $\sqrt{5}$ m s$^{-1} \approx 2.24$ m s^{-1}. The boatman has to row upstream at an angle α to the bank, where $\cos \alpha = \frac{2}{3}$; this gives $\alpha \approx 48°$.

(ii) In this case, the current is so strong that the boat will move downstream even if the boatman rows at full speed against the stream. This means, in contrast to the previous case, he cannot choose his direction with respect to the bank and, in particular, he cannot travel across in a direction perpendicular to the bank.

The possible directions he can take may be determined by adding all the possible still-water velocities of the boat to the velocity of the river. Draw the velocity vector of the river and, from the endpoint of this vector, draw velocity vectors in all directions, with a magnitude equal to the speed of the boat in still water. The endpoints of these vectors will form a circle as shown in Fig. S3.2.

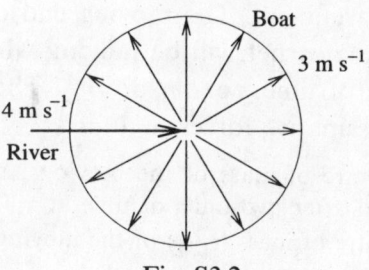

Fig. S3.2

The possible resultant velocities of the boat can be obtained by joining the starting point of the velocity vector of the river to the points on this circle. The resultant corresponding to the shortest path will be the one that makes the greatest angle with the direction of the current, i.e. when the line of action of the resultant velocity vector is a tangent to the circle (*see* Fig. S3.3).

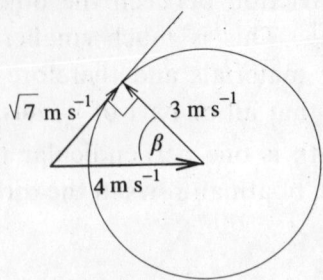

Fig. S3.3

Thus, the velocity of the boat with respect to the shore is $\sqrt{7}$ m s$^{-1} \approx$ 2.65 m s^{-1}. Again, the boatman has to row upstream, but this time at an angle β to the bank, where $\cos\beta = \frac{3}{4}$, yielding $\beta \approx 41°$. The figure also shows that in this case the distance travelled by the boat will be $\frac{4}{3}$ times the width of the river.

S4 Let the position of the moving end of the carpet be x as shown in the figure. It follows that the other end of the moving part is at $x/2$, and hence that the coordinate of its centre of mass is $3x/4$. Although $dx/dt = 1$, the speed of the centre of mass of the moving part is only $\frac{3}{4}$!

The linear momentum of the moving part is $p = mv$, where $v = 1$ and m is increasing uniformly with time. The net force acting on the moving part is thus

$$F = \frac{dp}{dt} = \frac{dm}{dt}v + \frac{dv}{dt}m = \frac{dm}{dt}1 + 0.$$

The rate of change of the mass of the moving part can be found with the help of the following argument. The moving end of the carpet starts from the origin and the whole carpet will be moving when it reaches $x = 2$; this it does after two units of time, i.e. $dm/dt = \frac{1}{2}$. The corresponding minimal force (neglecting all dissipative forces) is $F = \frac{1}{2}$.

> *Note.* (i) The centre of mass of the moving part of the carpet is initially at the origin and after two units of time at $x = \frac{3}{2}$, again showing that the speed of the centre of mass (v_{CM}) of the moving part is $\frac{3}{4}$.

(ii) Notice that the linear momentum of the moving part ($p = mv$) is *not* equal to the product (mv_{CM}) of its mass and the speed of its centre of mass.

(iii) It seems tempting to try to find the minimum force required by using the conservation of energy, i.e. $F \times 2L = mv^2/2$, where L is the length of the carpet, ($L = 1$). The result would be $F = \frac{1}{4}$, which is only one-half of the value calculated earlier. The error in this argument is to ignore the continuous inelastic collisions which occur when the moving part of the carpet is jerking the next piece into motion. Half of the work goes into the kinetic energy of the carpet, but the other half is dissipated as heat.

S5 *Solution 1.* If space (planar) and time coordinates are established in an orthogonal frame of reference with axes x, y and t (a space–time diagram), the 'world-line' of a snail travelling with uniform rectilinear motion will obviously be straight. Encounters occur when two snails are at the same place at the same time, i.e. when their world-lines intersect as shown in the figure.

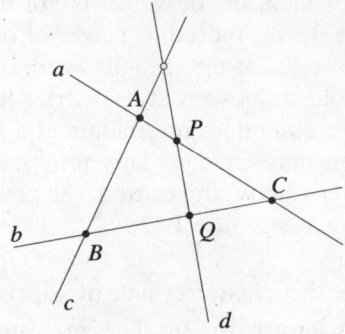

According to the information given in the problem, the four world-lines (a, b, c and d) definitely intersect in pairs at five 'points'. Let us denote these encounters by A, B, C, P and Q. Points A, B and C determine a plane (the plane of world lines a, b and c). Since P and Q also lie in this plane, world line d must do so as well. This means that world lines c and d also cross each other, implying that the sixth encounter will also take place.

Solution 2. As five encounters have already occurred, there has to be a snail who has already met all three of his fellows. Let us denote this snail by α. In our imagination, let us sit on the back of α, i.e. choose a frame of reference in which α is at rest at the origin.

The three other (moving) snails, (β, γ and δ) have already met α, and have therefore crossed the origin. Moreover, one of them (β say) has already met his two other moving fellows (since five encounters have occurred).

This is only possible if β, γ and δ are moving along the same straight line. Consequently, sooner or later, γ and δ will have to meet as well.

S6   The work done against gravity can be calculated from the increases in height of the centres of mass. The centre of mass of a worm 'folded in two' is located at the middle of either half, i.e. at a point one-quarter of the worm's total length from one end. This is illustrated in the figure.

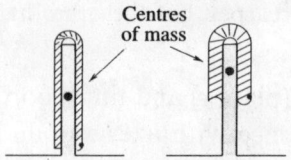

Thus, the centre of mass of the narrow flatworm travels 5 cm up the wall, whilst that of the broad one moves 7.5 cm. The ratio of the amounts of work done is therefore 2 : 3.

> *Note.* The centre of mass of the worm is not always in the same position with respect to the worm; indeed, it need not be at any point of the worm at all. The centre of mass of the straight worm is obviously at its centre, and that of the worm folded in two is at its quarter-length point. Thus the centre of mass of a flexible body does not remain at a fixed point within the body; its relative position may change. This principle is used by high-jumpers; when a high-jumper's body arches over the cross-bar, the body's centre of mass remains below it (*see also* P32).

S7   (i) Let us denote the elastic constant (spring constant) of the rope by k and its unstretched length by ℓ_0. The maximum length of the rope is $\ell_1 = h - h_0 = 23$ m, whilst in equilibrium it is $\ell_2 = (23 - 8)$ m $= 15$ m. Initially, and at the jumper's lowest position, the kinetic energy is zero. If we ignore the mass of the rope and assume that the jumper's centre of mass is half-way up his body, we can use conservation of energy to write

$$mgh = \frac{1}{2}k\left(\ell_1 - \ell_0\right)^2.$$

In addition, in equilibrium,

$$mg = k\left(\ell_2 - \ell_0\right).$$

Dividing the two equations by each other we obtain a quadratic equation for ℓ_0,

$$\ell_0^2 + 2(h - \ell_1)\ell_0 + (\ell_1^2 - 2h\ell_2) = \ell_0^2 + 4\ell_0 - 221 = 0,$$

which gives $\ell_0 = 13$ m.

(ii) When the falling jumper attains his highest speed, his acceleration must be zero, and so this must occur at the same level as the final equilibrium position ($\ell = \ell_2$).

Again applying the law of conservation of energy,

$$\frac{1}{2}mv^2 + \frac{1}{2}k(\ell_2 - \ell_0)^2 = mg(\ell_2 + h_0),$$

where the ratio m/k is the same as that obtained from the equilibrium condition, namely,

$$\frac{m}{k} = \frac{\ell_2 - \ell_0}{g}.$$

Substituting this into the energy equation, shows that the maximum speed of the jumper is $v = 18 \text{ m s}^{-1} \approx 65 \text{ km h}^{-1}$. It is easy to see that his maximum acceleration occurs at the lowest point of the jump. Since the largest extension of the rope (10 m) is five times that at the equilibrium position (2 m), the greatest tension in the rope is $5mg$. So the highest net force exerted on the jumper is $4mg$, and his maximum acceleration is $4g$.

S8   If the berg has base area A and height H, then $M = \frac{1}{3}AH\rho_{\text{ice}}$. If the height showing above the surface is h, the flotation condition gives $(H^3 - h^3)\rho_{\text{water}} = H^3\rho_{\text{ice}}$. When the berg is depressed by a small amount x the additional submerged volume is $xA(h/H)^2$ and the upthrust is this multiplied by $\rho_{\text{water}}g$. This gives that the angular frequency of oscillation ω is determined by

$$\omega^2 = \frac{3h^2\rho_{\text{water}}g}{\rho_{\text{ice}}H^3}$$

and, on substituting numerical values, that the period of oscillation is about 11 s.

S9   First of all we note that the right front suspension spring will be further compressed as a result of parking on the pavement. We can measure both the change of tensions in the suspension springs and the rise of the car body in centimetres, and will let the sign be positive if the spring is further compressed. The net torque must be zero about any axis, including, for example, the diagonals of the rectangle formed by the wheels; so the changes in tension at the opposite ends of a diagonal must be equal. This is why the springs of the right front (rf) and left back (lb) wheels are equally compressed, by an amount x, and the left front (lf) and right back (rb) suspension springs each lengthen by x. This equality of changes in length ensures that the net force provided by the springs to support the weight of the car does not change.

The rises of the car body both at the lf and rb wheels are x, at the lb wheel the rise is $-x$ and at the rf wheel (on the pavement) it is $8 - x$. The frame of the car is rigid, so, because of the equal movements at the lf and rb wheels, the midpoint of the chassis also rises by x. Similarly, the other chassis diagonal remains a straight line, and so the rise at the rf wheel must be the same as the fall of the body at the lb wheel relative to the midpoint of the chassis, i.e. $(8 - x) - x = x - (-x)$. From this very simple equation we get $x = 2$ cm. We conclude that above the wheel on the pavement the body of the car rises 6 cm, above the left back wheel it sinks 2 cm and above the other two wheels it rises 2 cm.

Applying the same calculational technique, it is easy to show that compressions in the suspension springs cannot change when the car is parked with both right wheels on the pavement. It follows that then the right side of the car body simply rises 8 cm, the height of the pavement. You can also show that the result does not depend upon the number and the positions of the people sitting in the car; this is because we have only investigated the *relative* displacement of the body of the car before and after parking on the pavement.

> *Note.* In the solution above, a slight rotation of the body of the car was ignored.

S10   Fig. S10.1 shows Jean Valjean's location on the wall. Figure S10.2 is a sketch showing his weight (mg), the normal reactions of the walls (N) and the static frictional forces (F_{fr}) acting on his limbs.

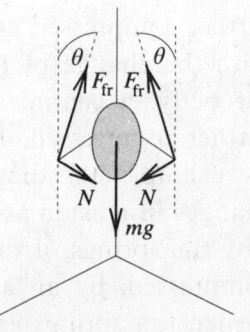

Fig. S10.1   Fig. S10.2

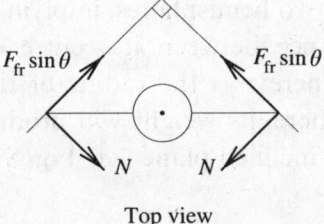

Fig. S10.3

Let the static frictional forces make a common angle θ with the vertical. The conditions for static equilibrium (*see* Fig. S10.3) are

$$mg = 2F_{\text{fr}}\cos\theta \quad \text{and} \quad N = F_{\text{fr}}\sin\theta.$$

From these equations we can express the normal component, N, of the force exerted by the prisoner on the wall whilst climbing as

$$N = \tfrac{1}{2}mg\,\tan\theta.$$

Thus the total force required, F, is given by

$$F^2 = N^2 + F_{\text{fr}}^2 = \left(\frac{mg}{2}\right)^2 \frac{1+\sin^2\theta}{\cos^2\theta}.$$

We can also find the minimal force using the inequality

$$F_{\text{fr}} \leq \mu_0 N,$$

from which it follows that

$$\sin\theta \geq \frac{1}{\mu_0} \quad \text{or} \quad \tan\theta \geq \frac{1}{\sqrt{\mu_0^2-1}},$$

where μ_0 is the coefficient of static friction. Using either of these inequalities we find the minimal force to be

$$F_{\min} = \frac{mg}{2}\sqrt{\frac{\mu_0^2+1}{\mu_0^2-1}}.$$

This expression shows that the coefficient of static friction must be greater than unity if Jean Valjean is not to fall off the wall. If the coefficient of static friction approaches infinity, the force on each of his hands is equal to half of his body weight; this situation corresponds to his being glued to the wall.

S11 If static friction is large enough, the sphere will not slide down the slope. However, this by itself is not sufficient for equilibrium; it is also necessary that the sphere does not *roll* down the inclined plane.

The sphere is made of two hemispheres, implying an inhomogeneous mass distribution. If the distance between its centre of mass and geometrical centre is less than $\frac{1}{2}r$, where r is the radius of the sphere, then, whatever the orientation of the sphere, its weight will produce a torque about P, the point of contact with the inclined plane (*see* Fig. S11.1), which will make the sphere roll.

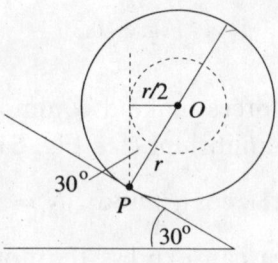

Fig. S11.1

It will now be shown that this is the situation for any sphere made of two homogeneous hemispheres – whatever the densities of the two halves.

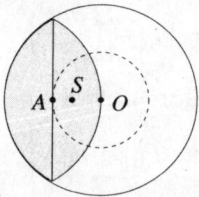

Fig. S11.2

Consider the shaded area in Fig. S11.2. By symmetry, the centre of mass of this part is obviously at point A, i.e. at a distance $\frac{1}{2}r$ from the centre, O. The rest of the sphere moves the centre of mass S of the whole even closer to point O, i.e. $OS < \frac{1}{2}r$. From our previous considerations, this implies that the sphere cannot remain in equilibrium on the 30° inclined plane. In obtaining the solution, we have assumed that rolling resistance is small, i.e. no resistant torque can act at point P. In the case of a surface covered with Velcro, this is obviously not true, and the sphere may even adhere to a vertical surface.

S12 Viewed within a rectangular coordinate system which has one axis parallel to the inclined plane, the ball is seen to bounce on a 'horizontal' plane in a 'vertical' field of gravitational acceleration $g' = g \cos \alpha$. It also experiences an additional constant 'horizontal' acceleration (of magnitude $g \sin \alpha$). The 'vertical' motion consists of bounces of identical heights, i.e.

of identical periods. Meanwhile, since the 'horizontal' acceleration is constant, the ball's average speed between bounces increases uniformly, and so the distances between two consecutive bounces increase in an arithmetical progression.

S13 Let the midpoint of the platform be a distance h below the pivot and the hamster's distance from that midpoint be x, as shown in the figure.

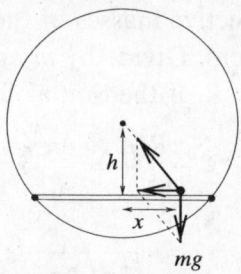

Because of gravity the hamster exerts a torque mgx about the pivot of the wheel-cage. On the other hand, as the hamster moves it accelerates using friction with the platform. When its acceleration is a this produces a reaction force of ma on the platform, directed away from its midpoint. The torque due to this force is mah. The wheel-cage (and the platform) remains in static equilibrium if these two torques are equal, i.e.

$$mgx = mah.$$

After making due allowance for its direction, the acceleration can thus be written as $a = -(g/h)\,x$. This shows that the required motion of the hamster is simple harmonic motion with an angular frequency $\omega = \sqrt{g/h}$.

S14 (i) The bicycle moves in the direction of the *net* force (the sum of the applied backward force and the frictional force directed forwards). In usual gearings the bike moves backwards, but extremely low gearings can cause forward displacement. Because the work done is always positive, it follows that the student's hand moves backwards relative to the ground. Normally the gearing N is greater than one, i.e. the rear wheel rotates more rapidly than the pedals. However, in the unusual case $N < r/R < 1$ (where R is the radius of the wheel and r is the length of the pedal arm), the bicycle could move forwards despite the oppositely applied force.

(ii) The chain-wheel rotates in the same sense as the rear wheel.

(iii) Usually backwards and upwards; the superposition of (i) and (ii), but with (i) larger because of the gearing and wheel sizes.

> *Note.* It is interesting to note that there is one position of the pedals at which an arbitrarily large force can be applied without moving the bicycle either way.

S15   Equate the force of attraction between the Sun and the Earth to the centripetal force that keeps the Earth in its approximately circular orbit, and express the angular velocity ω in terms of T, the period of revolution. This gives

$$G\frac{mM}{r^2} = mr\omega^2 = mr\frac{4\pi^2}{T^2},$$

where m and M are the respective masses of the Earth and Sun, and r is the average distance between them. Divide by m and express M in terms of the average density ρ and radius R of the Sun as follows:

$$G\frac{\frac{4}{3}\pi R^3 \rho}{r^2} = \frac{4\pi^2}{T^2}r.$$

This yields

$$T = \sqrt{\frac{3\pi}{G\rho}\left(\frac{r}{R}\right)^3}$$

for the period of revolution.

It can be seen that the Earth's rotation period only depends on the universal gravitational constant G, the average density of the Sun and the ratio r/R. Therefore if the density of matter remains constant, any scaling of the solar system leaves the length of a year unchanged. It can also be seen that only the density and size of the Sun are relevant; the Earth's data are not. Any body that is small in size relative to the Sun would have the same period and follow the same orbit.

> *Note.* This result can also be obtained using Kepler's third law $T^2/a^3 = 4\pi^2/GM$, where a is the semi-major axis of the Earth's elliptical orbit. If the mass of the Sun is expressed in terms of its average density then it is clear that a proportional reduction does not change the period of planets in elliptical orbits.

S16   The gravitational acceleration produced by the Sun at the position of the Earth is the same as that due to one of the stars at the position of the other. This is because, according to Newton's law of gravitation, g (the gravitational force acting on a unit mass) depends only on the mass at the centre of attraction and the distance of the second body from it. These quantities are identical in the two systems.

Thus, the members of the binary star move with the same acceleration as the Earth but in an orbit of radius only half that of the Earth's orbit. This means that, since $(a = r\omega^2)$, the square of their angular velocity has to be twice as large as that of the Earth. The period of the binary star therefore equals that of the Earth around the Sun divided by $\sqrt{2}$, i.e. $8\frac{1}{2}$ months.

S17 (i) The acceleration of a satellite moving at speed v in a circular orbit of radius R is $g = v^2/R$. If R is the radius of the Earth (or more precisely, a slightly larger value), then g has to be the gravitational acceleration at the Earth's surface; this defines the 'first cosmic speed', $v_1 = \sqrt{Rg} = 7.9$ km s^{-1} for the speed of the satellite.

All of this speed is not strictly necessary for launching a satellite if the 'initial speed' provided by the rotation of the Earth is taken into account. This help is greatest at the Equator, approximately 0.5 km s^{-1}, and means that an initial speed of 7.4 km s^{-1} with respect to the Earth is sufficient, but only if the satellite is launched *eastward* from on or near the Equator.

(ii) The angular momentum of satellites in polar orbits (passing over the poles) is zero with respect to the axis of rotation of the Earth. This condition has to be fulfilled right from the launch, since the angular momentum will not change later. The 'help' described above cannot therefore be utilised. Indeed, the rotation of the Earth is a drawback, since, not only must the satellite receive the speed of 7.9 km s^{-1} in a north–south direction, but, in addition, the unacceptable west–east speed due to the rotation of the Earth has to be cancelled. The latter would not have to be taken into account if the satellite were launched from the neighbourhood of one of the poles; there are obvious technical difficulties in doing this!

The initial speed necessary to put a satellite into a polar orbit is therefore at least $7.9/7.4 \approx 1.06$ times greater, i.e. the necessary kinetic energy is at least 1.13 times as great. This does not seem a big difference, but in reality the slightest increase of the initial speed requires an enormous effort. The reason for this is that the carrier rocket has to be accelerated as well as the satellite, and the mass to be launched increases exponentially with the intended final speed.

(iii) In order to escape the attraction of the Earth (i.e. to move far from the Earth), the probe has to acquire the escape or 'second cosmic' speed, $v_2 = \sqrt{2gR} = \sqrt{2}v_1 \approx 11.2$ km s^{-1}. The rotation of the Earth can again be used. Launching eastward from the Equator, a launch speed of 10.7 km s^{-1} relative to the Earth is sufficient.

(iv) The Earth revolves around the Sun at a speed of approximately 30 km s^{-1}. In order to reach the Sun, a space probe has to be launched at an initial speed of 30 km s^{-1} (or more exactly, 29.5 km s^{-1}; smaller by 0.5 km s^{-1} due to the rotation of the Earth). If the aim is to leave the solar system, the space probe has to reach $\sqrt{2}$ times the speed of 30 km s^{-1}, but the 'initial speed' of the Earth in its orbit can be subtracted from this value if the launching is well timed and directed, i.e. a speed of 12 km s^{-1} is sufficient.

Thus, it is easier to make a space probe leave the solar system than to send it into the Sun. (The former has been successfully attempted, the latter we are still waiting for.) The situation is even more favourable if the possibilities offered by the outer planets (Mars, Jupiter, Saturn, ...) are taken into account. A space probe launched at the right time and in the right direction can be significantly 'pushed by' (i.e. receive energy from) these planets, a phenomenon known as 'gravitational slingshot'. Designed in this way, the space probe does not have to propel itself very far, it only has to reach Mars or Jupiter and the rest happens 'automatically'.

S18   The rocket has to reach the highest possible total energy. If the zero level of gravitational potential energy is 'infinitely' far away, then the energy of the rocket standing on the surface of the Earth is negative. The energy released during the operation of the engines increases the total energy of the rocket, and the rocket can leave the Earth's gravitational field if the sum of its potential and kinetic energies becomes positive.

The energy released in the course of the operation of the principal and auxiliary engines increases the total energy of the rocket and its ejected combustion products by a fixed value; this increase is independent of the moment when the engines are switched on. However, the speed at which the combustion products fall back to the Earth does depend on the timing of the rocket's operations. Indeed, if the auxiliary engine starts working when the rocket is at a greater height, the combustion products fall further and their speed and total energy are higher when they hit the ground. This means that the sooner the auxiliary engine is switched on, the higher the energy ultimately acquired by the rocket. The same argument is valid for the principal engine, and if only energy considerations apply, it is best to operate the engines for the shortest time and at the highest thrust.

S19   The mass distribution, and thus the position of the centre of mass, changes from moment to moment as the ball sinks. In a time t, the centre of mass is displaced by

$$s = vt \frac{\rho_b V - \rho_h V}{M},$$

where M is the total mass of the system, V the volume of the ball, and ρ_h and ρ_b are the respective densities of the honey and the ball. This is so because when the ball has moved through a distance vt, it can be considered to have changed places with a 'honey ball' of identical volume. Thus the total momentum of the system is

$$p_{total} = \frac{Ms}{t} = v\rho_b V - v\rho_h V.$$

The first term on the right-hand side is the momentum of the steel ball, and therefore the second is that of the honey:

$$p_{\text{honey}} = -v\rho_{\text{h}}V = -2 \text{ g cm s}^{-1}.$$

The negative sign shows that the direction of the honey's momentum is upwards. Its magnitude is the same as that of a honey ball moving upwards with a speed equal but opposite to that of the steel ball.

S20 The (average) kinetic energy of the gas molecules is proportional to the square of their velocity. The internal energy of the gas is proportional to the temperature. Therefore $v^2 \sim T$. If the wall is warmer than the gas ($T_1 > T$) then the average speed of the rebounding gas molecules will be increased by the collision (the wall warms the gas). If the wall is colder than the gas ($T_1 < T$) then the situation is reversed; the molecules rebound with a lower speed (the gas cools down).

From a molecular point of view, gases exert a pressure on the walls of their container because of the changing momentum of molecules that hit the wall and rebound from it. For a given initial momentum and collision rate, the rate of change in the momentum of molecules rebounding from a warm wall is greater than that of molecules rebounding from a cold one. Thus, the gas exerts a higher pressure on a warm wall than on a cold one.

> *Note.* This phenomenon explains the unexpected rotation of a radiometer ('light wheel'). If one side of each blade of a wheel mounted on a delicate bearing is black and the other one is shiny, then the wheel starts to turn when it is illuminated. At first sight, one might be tempted to think that the pressure associated with the reflection of the light turns the wheel. This, however, is not true, since experience shows it is, in fact, the shiny side (the one reflecting rather than absorbing the light, and hence causing the greater change in photon momentum) of the blades that moves forwards! The correct explanation is that the black side of the blades warms up more and therefore the pressure of the air molecules rebounding from that side is greater than on the colder shiny side.

S21 As a result of thermal expansion, the size of both spheres increases. The centre of mass of the sphere lying on the plate rises, whilst that of the sphere hanging on the thread sinks. Thus, the potential energy of the first sphere increases, whilst that of the second one decreases as shown in the figure.

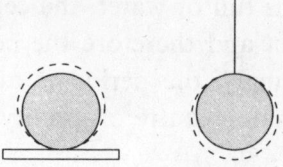

According to the first law of thermodynamics, the heat transferred to the spheres produces not only an increase in internal energy and the small amount of work done in expanding against the atmospheric pressure (this is the same for both spheres), but also a change in gravitational potential energy. The potential energy of the sphere lying on the insulating plate increases a little, therefore its internal energy increases by less than the residual heat transferred. Conversely, the decrease in potential energy of the hanging sphere contributes positively to the increase in its internal energy. In summary, the temperature of the sphere suspended from the thread will be higher.

It is worth giving a numerical estimate. If the temperature of the two iron balls, each with a radius of 10 cm, is increased by 100 °C, a temperature difference of $\Delta T \approx 5 \times 10^{-6}$ °C will result from this effect. This is undetectable in practice.

S22   For quantitative purposes we assume that the resistances of the bulbs do not depend upon the voltages across them. This is far from accurate, but will give the correct qualitative conclusion. If the (r.m.s.) supply voltage is V, the resistance r_i of a bulb is V^2/w_i, where w_i is the nominal rating of the bulb. When the two bulbs are connected in series across the supply, the (r.m.s.) current drawn is $V/(r_A + r_B)$ and the power dissipated in bulb i ($i = A$ or B) is

$$P_i = \frac{V^2}{w_i} \left[\frac{V}{(V^2/w_A) + (V^2/w_B)} \right]^2 .$$

According to the original agreement ($w_A = w_B = 100$ W), both P_A and P_B should be 25 W. Actually, $P_A = 8$ W and $P_B = 32$ W, and so A clearly failed his examinations. By comparison, student B might be considered a double winner: he gets 32 W, but pays for only $(8 + 32)/2 = 20$ W. On the other hand, 32 W is still a very poor light to study by and B also could well have failed his examinations.

S23   A simple circuit consisting of two identical resistors connected as shown in the figure would behave as described.

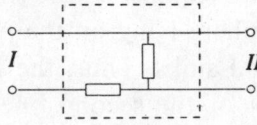

S24   When the bucket is full of water, the centre of gravity of the water is above that of the bucket and therefore the common centre of gravity is at its highest; correspondingly, the period is at its shortest. As the water starts leaking out, the common centre of gravity moves downwards and the period becomes longer. When the bucket is half full, the centre of gravity of the water is below that of the bucket and the common centre of gravity has moved even lower, significantly lower than the centre of gravity of the bucket. Thus the period has increased further. When there is no water left in the bucket, the centre of gravity coincides with that of the empty bucket, which is higher than in the previous cases, i.e. the common centre of gravity stopped moving downwards at some point and started to move upwards again. In summary: the longest period occurs when the common centre of gravity is at its lowest position. As is shown in the next problem, this occurs when the common centre of gravity lies in the water surface.

S25 Clearly, the first water to be added is placed below the centre of gravity of the empty beaker and therefore lowers the overall centre of gravity. If at some stage water is added above the current overall centre of gravity, the latter will be raised. Therefore for maximum stability the overall centre of gravity must lie in the water surface.

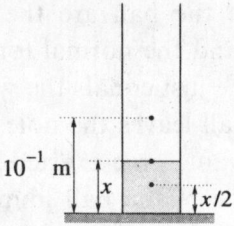

Thus

$$10^{-1} \times 10^{-1} + \pi \left(3 \times 10^{-2}\right)^2 10^3 x \frac{x}{2} = x \left[10^{-1} + \pi \left(3 \times 10^{-2}\right)^2 10^3 x\right],$$

giving $x = 55.9$ mm.

S26 Assume first that the centre of mass of the bowl remains at the same height as originally. Then the bowl only turns about its centre (and perhaps moves sideways) but does not sink any deeper into the water. Under these circumstances, the rim of the bowl is lowered to the water surface on the side opposite the chain, and water flows into the bowl. We will now prove that this cannot occur.

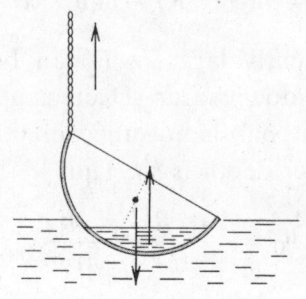

The upthrust acting on the body in the assumed situation remains the same as in the initial one, i.e. it equals the total weight of the bowl and soup. Thus, the chain cannot be exerting any force on the bowl. On the other hand, the centre of mass of the bowl is not on the line of action of the upthrust when the edge of the bowl is being raised, and the torques can only be balanced if the chain does exert a force and pulls the bowl upwards.

The two contradictory conditions show that our initial assumption was wrong. The geometrical centre of the bowl cannot stay in the same place but has to rise (since a smaller upthrust is sufficient when the chain exerts an upward force). This implies that even the lowest point of the rim of the bowl has to remain above the surface of the water.

The possibility of the soup flowing out into the water has also to be considered. This could occur if the level of the soup in the bowl were higher than the water level in the lake, i.e. if the density of the soup were lower than that of water. Realistically this would not be the case.

S27 The forces acting on the ball are the gravitational force mg, the buoyancy force of the water and the normal force exerted by the rim of the hole. When the buoyancy force just equals the weight of the ball, the normal force becomes zero and the ball leaves the hole.

We first calculate the buoyant force exerted on the ball when the water depth is h. Denote the volume of the ball immersed in water by V, where $V = V(r, R)$. Now, imagine the ball to have that part of it which protrudes through the hole removed and the space under the container filled with water. The buoyancy force would then be

$$F_1 = \rho g\, V(r, R),$$

where ρ is the density of water. As in reality there is no water under the hole, a contribution

$$F_2 = \rho g h \times \pi r^2$$

is missing. Thus the actual buoyancy force exerted on the ball is

$$F = \rho g\, V(r, R) - \rho g h \times \pi r^2.$$

It is clear that, for sufficiently large h, F can be negative and then the 'buoyant' force is directed downwards. Decreasing h causes F to increase, and, provided the top of the ball is not uncovered, it will rise when $F = mg$ and the corresponding water depth is h_0. Thus

$$h_0 = \frac{V(r, R)}{\pi r^2} - \frac{m}{\pi r^2 \rho}.$$

Straightforward integration, or geometry books, give the volume of a spherical calotte (truncated sphere) as

$$V(r, R) = \frac{\pi}{3}\left[2R^3 + \left(2R^2 + r^2\right)\sqrt{R^2 - r^2}\right],$$

and using this formula we can express h_0 as

$$h_0 = \frac{2R^3}{3r^2} + \frac{2R^2 + r^2}{3r^2}\sqrt{R^2 - r^2} - \frac{m}{r^2\pi\rho}.$$

Don't forget that the formula above is only valid if the top of the ball is still covered by water, that is $h > R + \sqrt{R^2 - r^2}$. Otherwise we have to modify $V(r, R)$; instead of a spherical calotte we have to calculate the volume of a sphere truncated at both ends of a diameter. This leads to a cubic equation for the critical h_0,

$$h_0^2\left(3\sqrt{R^2 - r^2} - h_0\right) = \frac{3m}{\pi\rho}.$$

Note. When calculating the buoyancy force F, instead of using the radius of the hole r we can work with another variable: $\ell = \sqrt{R^2 - r^2}$ (*see* Fig. S27.1).

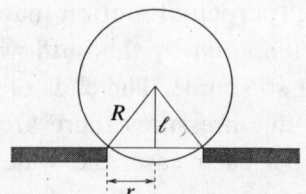

Fig. S27.1

Either by applying the previous argument or by integrating the vertical component of the upthrust over the submerged part of the sphere, we can calculate the buoyancy force to be

$$F(h) = \begin{cases} \dfrac{3\ell^2(h - \ell) + 2\ell^3 - (h - \ell)^3}{3}\pi\rho g, & \text{if } h \leq R + \ell, \\[3mm] \dfrac{2\left(R^3 + \ell^3\right) - 3(h - \ell)\left(R^2 - \ell^2\right)}{3}\pi\rho g, & \text{if } h \geq R + \ell. \end{cases}$$

The expression for the buoyancy force F as a function of h consists of two parts; in the interval $0 \leq h \leq R + \ell$ the force is a cubic function of h, whilst in the range $h \geq R + \ell$ it decreases linearly, as shown in Fig. S27.2.

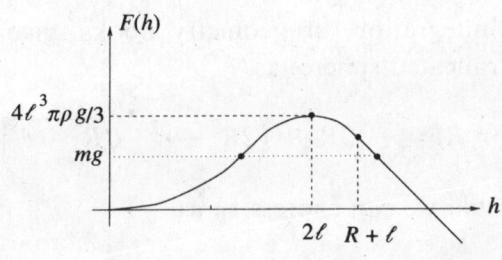

Fig. S27.2

We can find the critical water depth h_0 from the intersection of the graph of $F(h)$ with the line $F = mg$. Differentiating $F(h)$ shows that it has a maximum value of $(4\ell^3\pi\rho g/3)$ at $h = 2\ell$. If the weight of the ball is larger than this value (which is just the weight of a water sphere of radius ℓ), then the ball will not float out of the hole, whatever depth of water we have.

S28   The soap bubble floats and therefore the combined mass of its wall and the helium inside it is equal to that of the displaced air. Since the density of helium is less than half the density of air, the mass of the helium is less than half of the mass of the displaced air. Thus, the wall of the bubble has to be heavier than the gas it encloses.

S29   In case (*a*), it is clear that the water cannot flow out of the tube. If it could, a perpetuum mobile (perpetual motion machine) could be established using a paddle rotated *ad infinitum* by the outflowing water.

Cases (*b*) and (*c*) are not so simple. The ends of both tubes are lower than the water level and the water pressures there are consequently lower than the atmospheric pressure. In each case, the water wells to such an extent that the pressure corresponding to its radius of curvature equals the pressure difference between it and the air. The water surfaces corresponding to cases (*a*), (*b*) and (*c*) are shown in the figure.

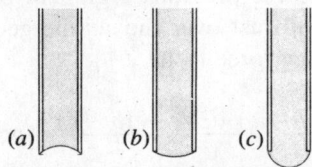

The greatest curvature (smallest radius of curvature) occurs for the hemispherical shape and corresponds to the pressure of a column of water of height H, since this is the height reached by the water in a vertical tube. If the air–water pressure difference is greater than $\rho g H$, then surface tension cannot hold the water in the tube and it flows out. This is what happens in tube (c) (assuming that the figure is to scale and $H' > H$); on the other hand, water does not flow out of tube (b).

S30 The system described in the problem is spherically symmetrical. Therefore the magnetic field that is built up has to be spherically symmetrical as well. A spherically symmetrical vector field has to be radial everywhere and its magnitude can depend only on the distance from the origin: $\mathbf{B}(\mathbf{r}) = B(r)\,\mathbf{r}/|\mathbf{r}|$.

On the other hand, a magnetic field contains no sources (magnetic monopoles) and the magnetic flux crossing any closed surface has to be zero at any given moment. In particular, consider a spherical surface of radius r around the capacitor. The consequences of sourcelessness can only be met if $B(r) = 0$ for any r. This means that the current described in the problem builds up no magnetic field, either inside or outside the spherical capacitor.

> *Note.* It is worth examining how the basic laws of electrodynamics are satisfied between the plates of the spherical capacitor. Is it true that a magnetic field builds up around a current flowing in a conductor, and that the rotation (curl) of this field is proportional to the current?

S31 The radiation has to be spherically symmetrical, since both the distribution and the motion of the charges are spherically symmetrical. The magnetic field is always radial and should have the same magnitude at any given distance from the sphere, irrespective of direction. This, however, is impossible, since such a magnetic field (unless of zero magnitude) would imply the presence of a magnetic charge (magnetic monopole), something that experimentally is found not to exist in nature. Similar reasoning shows that the electric field is also spherically symmetrical and that its magnitude depends only on the total surface charge of the sphere and not on the pulsation parameters. Therefore, only the static Coulomb field can be observed outside the sphere, and the sphere *emits no radiation at all*!

If individual parts of the sphere are examined, they are found to behave like dipole antennae and emit radiation. But the radiation from different parts has to be summed, taking into account phases as well as magnitudes. The individual radiations from the many dipole antennae cancel each other out, a result which can be proved by direct (but lengthy) calculation without referring to spherical symmetry.

S32 No high-jump competition has yet been held on the Moon. However, here we try to estimate the expected result. The men's high-jump record on Earth is over 240 cm. A good male high-jumper is more than 190 cm tall and has a mass of around 80 kg, with his centre of mass approximately

110 cm above the ground. For a successful jump, all of his body must rise to the height of the cross-bar, but his centre of mass *need not*. This achievement requires a special jumping technique (the Fosbury flop), which can be studied in slow-motion video recordings. The centre of mass of the high-jumper remains approximately 20 cm below the cross-bar, even when he is at the apex of his jump. For the western roll and straddle jumping techniques, the jumper's centre of mass has to rise above the bar.

The most difficult part of our estimation is comparing the movement of a jumper at take-off on the Earth to that of a jumper leaving the ground on the Moon. Assume that the centre of mass of the high-jumper rises by $s = 40$ cm from its lowest point (in the crouch just before the jump) to the highest point (when he has just left the ground) both on Earth and on the Moon. Then his muscles must have done enough work to subsequently carry his centre of mass from 110 to 220 cm, i.e. raise it by $h = 110$ cm on Earth. Any effect of the run-up has been ignored, or has been assumed to be identical in both places.

The basis of our estimate is the assumption that the *same work* is done and the same jumping technique is employed in both cases. The work done is the sum of the kinetic energy of his body and the potential energy of his centre of mass, $W = \frac{1}{2}mv^2 + mgs$. His speed when leaving the ground can be calculated using the relation $v^2 = 2gh$. Thus, the total work done is

$$W = mg(h + s) = 80 \text{ kg} \times 10 \text{ m s}^{-2} \times 1.5 \text{ m} = 1200 \text{ J}.$$

We assume that the work done by the high-jumper on the Moon is the same, and that the rise s of his centre of mass before leaving the ground is also unaltered. The gravitational acceleration on the Moon is only one sixth of that on Earth, i.e. the energy equation of the jump on the Moon is $1200 \text{ J} = \frac{1}{6} mg (s + h')$, which yields $h' = 8.6$ m. This is the vertical height by which the jumper's centre of mass rises on the Moon. To this must be added the initial height of his centre of mass, 110 cm, and the extra 0.2 m resulting from his special technique, to give an estimated record of 9.9 m $\approx$ 10 m.

> *Note.* This question is usually answered – recognising only the difference in the gravitational accelerations – by saying that the world record on the Moon would be six times that on the Earth, i.e. approximately 14–15 metres. According to the above analysis this is an optimistic expectation. Even our estimate is probably over-optimistic, since in our model the high-jumper has to do the same amount of work in a shorter time, i.e. a greater power output is needed. It can be shown that the jumper has to increase his power output by nearly 15 per cent when on the Moon. If, instead, the speed at which he leaves the ground is taken to be the same as on Earth, a result

of 8 m is obtained; this will be too low a value. All things considered, the most probable value seems to be a height of about 9 m.

S33 (i) The vertical acceleration of ball B falling from the table is always g, which makes it possible to determine the time (approximately half a second) it takes to fall 1 m. The motion of the bob of the simple pendulum is rather complicated, as no small amplitude approximation is possible, and therefore the time during which it is in motion is not easy to determine. What can be stated with certainty is that, since the thread exerts an upward force on it, its vertical acceleration is always less than g. Therefore the vertical motion of ball A takes a longer time than the vertical free fall of ball B. Ball A stays in motion for longer.

(ii) The bob of the pendulum describes one-quarter of a circle (a path of approximately 1.5 m). The other ball, B, follows a parabolic path, the length of which cannot be determined by elementary methods. However, it is easy to see that it hits the ground at a distance of $vt = \sqrt{2 \times g \times 1}\sqrt{2 \times 1/g} = 2$ m from the edge of the table. The length of its path is therefore not less than the shortest distance between the beginning and end points of its motion, namely $\sqrt{5}$ m ≈ 2.2 m.

In summary: ball B moves on a longer path, but in a shorter time, than the bob of the pendulum.

S34 For uniform circular motion the tangential acceleration of the body is zero, but the radial acceleration is v^2/R, where v is the speed of the body and R is the radius of the circle.

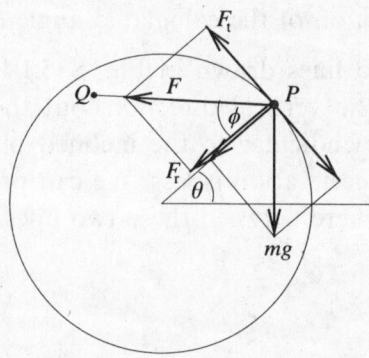

Fig. S34.1

As the radial and tangential components of the gravitational force are $mg\sin\theta$ and $mg\cos\theta$, respectively, as shown in Fig. S34.1, the force F

exerted by the string PQ on the bob must have components:

$$F_t = mg \cos \theta \qquad \text{and} \qquad F_r = \frac{mv^2}{R} - mg \sin \theta.$$

From these expressions we can deduce the direction of the force F, and hence the direction of the string, characterised by ϕ:

$$\cot \phi = \frac{F_r}{F_t} = \frac{v^2}{gR \cos \theta} - \tan \theta = \frac{1.83}{\cos \theta} - \tan \theta.$$

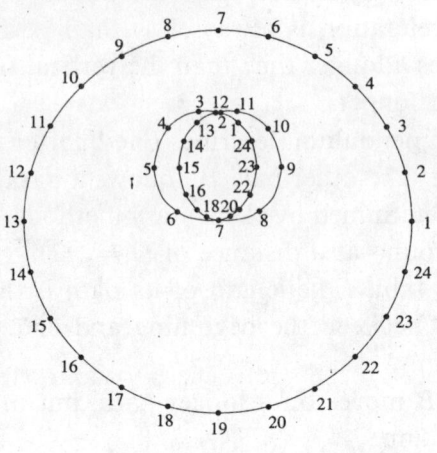

Fig. S34.2

In Fig. S34.2 corresponding positions of the two ends of the string are shown. The angle θ increases in 15° steps and the position of the other end (Q) of the string has been plotted from point P by marking off the length of the string in the direction of the calculated angle ϕ.

S35   Look at the two lines drawn in Fig. S35.1. The acceleration (g) is greater for the one in the vertical direction, but the path length involved is longer. The path perpendicular to the inclined plane is the shorter one, but the corresponding acceleration is less. We can presume that the path of shortest time lies somewhere between these two lines.

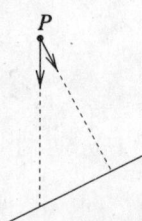

Fig. S35.1

We next prove the following auxiliary theorem: bodies starting at the same time $t = 0$, from the same point, and following frictionless slopes in different directions, all lie on a circle at any subsequent time.

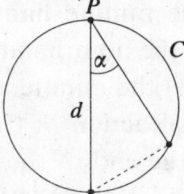

Fig. S35.2

As shown in Fig. S35.2, the topmost point of any such circle C is the starting point P. After time t, a body following a vertical wire and in free fall will have fallen through $d = \frac{1}{2}gt^2$, and this must be the diameter of C. A body moving along a wire at an angle α to the vertical has an acceleration of $g \cos \alpha$. In the same time t it will have covered a distance, measured from P, of $\frac{1}{2}g \cos \alpha\, t^2 = d \cos \alpha$. But this is precisely the length of the chord of C cut off by the wire. Thus, independent of α, the second body also lies on C – and the auxiliary theorem is proved.

The original problem is easily solved using the auxiliary theorem. Bodies starting at the same time from point P and travelling in different directions, always form a circle that grows with time and has P as its topmost point. After some time, the circle will touch the inclined plane, with the plane tangential to the circle at the contact point P'. A body starting from point P reaches the plane in the shortest time by travelling along the line PP'. In fact, the problem is three-dimensional, and bodies starting from point P lie on a sphere at any one time. The shortest time direction is found by joining P to the point of the sphere that first touches the inclined plane. However, it is sufficient for the question in hand to examine the vertical cross-section through P parallel to the plane's line of greatest slope, as we have done so far.

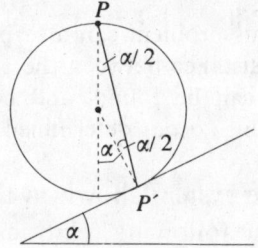

Fig. S35.3

It is clear from Fig. S35.3 that in the case of a plane inclined at angle α to the horizontal, the line PP' corresponding to the shortest time makes an angle $\alpha/2$ with the vertical, i.e. the optimium direction bisects those of the two lines mentioned in the first paragraph of the solution.

S36   The solution to the problem is surprisingly easy using a rotating frame of reference fixed to the minute hand. In this reference frame the minute hand is at rest, whereas the hour hand is moving 'anti-clockwise'. The separation between (the ends of) the minute and the hour hands increases at the highest rate when the line of action of the velocity vector of the end of the hour hand passes through the end of the minute hand. In this situation, the two hands and the line joining their ends form a right-angled triangle, as shown in the figure.

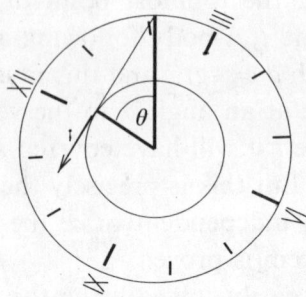

Since the minute hand is twice as long as the hour hand, the angle between the hands will be $\theta = \cos^{-1}(1/2) = \pi/3$. We can now calculate the exact time after midnight when the angle between the hands is θ. As the minute hand moves 12 times as fast as the hour hand, the angle ϕ between the hour hand and the 12 o'clock position is given by $12\phi - \phi = \theta$, i.e. $\phi = \frac{1}{11}\theta$. Thus, since midnight, the minute hand has moved through an angle of $\frac{12}{11}\theta = \frac{4}{11}\pi$ and the time is just before 11 minutes past midnight. There are several subsequent times (twice in each hour) when the angle between the hands is the same. The second occurs when the ends of the hands approach each other at the fastest rate.

> *Note.* Using calculus this problem can also be solved by brute force. From an expression for the distance between the ends of the hands of the clock, their relative velocity can be found, and hence the angular positions at which its stationary values occur determined.

S37   Using the coordinate system shown in the figure, the motion of the stone can be described by the following relations:

$$x = v_0 t \cos\alpha, \quad y = v_0 t \sin\alpha - \frac{g}{2}t^2,$$

$$v_x = v_0 \cos\alpha, \quad v_y = v_0 \sin\alpha - gt.$$

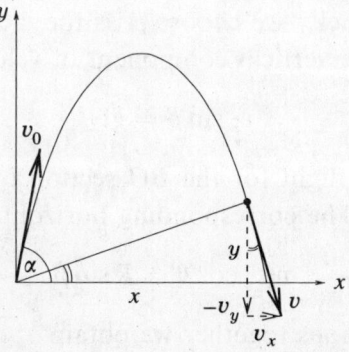

The stone is at the greatest distance from the origin when its velocity is perpendicular to its position vector. The condition for this is

$$\frac{y}{x} = -\frac{v_x}{v_y},$$

which yields a quadratic equation for the time t at which this happens:

$$t^2 - \frac{3v_0 \sin \alpha}{g} t + \frac{2v_0^2}{g^2} = 0.$$

If this is not to happen, the discriminant of this equation must be negative, i.e.

$$\left(\frac{3v_0 \sin \alpha}{g} \right)^2 < 4 \left(\frac{2v_0^2}{g^2} \right).$$

Thus, for the stone to be permanently moving away from the thrower, we must have $\sin \alpha < \sqrt{8/9} = 0.94$, i.e. $\alpha < 70.5°$.

S38 The trajectory of the grasshopper is a parabola, which touches the trunk at two symmetrically placed points, B and B^*, on the two sides of the trunk (at the moment we don't know anything about these points – they may or may not coincide at the topmost point E of the trunk). The grasshopper takes off from point A with an initial speed v_1 and at an angle θ with the horizontal, as shown in the figure. At the tangential points B and B^* the grasshopper's velocity is v_2, making an angle β with the horizontal.

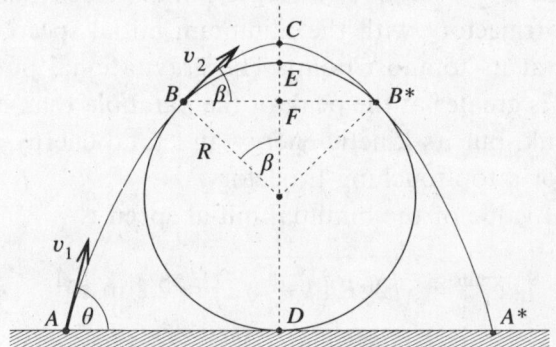

For the sake of simplicity we choose β as the independent variable of the problem. At point B the vertical component of velocity is

$$v_2 \sin \beta = gt_2,$$

where t_2 is the time of flight for the BC section of the trajectory (C is the peak of the parabola). The corresponding horizontal displacement BF is

$$v_2 t_2 \cos \beta = R \sin \beta.$$

Multiplying these equations together we obtain

$$v_2^2 = \frac{gR}{\cos \beta}.$$

Conservation of energy between points A and B of the trajectory gives

$$\frac{1}{2}mv_1^2 = \frac{1}{2}mv_2^2 + mg(R + R\cos \beta),$$

and so

$$
\begin{aligned}
v_1^2 &= v_2^2 + 2gR(1 + \cos \beta) \\
&= \frac{gR}{\cos \beta} + 2gR(1 + \cos \beta) \\
&= 2gR \left(1 + \cos \beta + \frac{1}{2\cos \beta} \right).
\end{aligned}
$$

We can calculate the minimum value of v_1 using differential calculus. However, there is a less complicated method available which uses the inequality between arithmetic and geometric means:

$$\frac{1}{2} \left(\cos \beta + \frac{1}{2\cos \beta} \right) \geq \sqrt{\cos \beta \frac{1}{2\cos \beta}} = \frac{\sqrt{2}}{2}.$$

So the minimum value of $\cos \beta + 1/(2\cos \beta)$ is equal to $\sqrt{2}$ and, therefore, $\beta = 45°$. The case $\beta = 0$ requires a larger initial velocity, since $1.5 > \sqrt{2}$; it follows that the trajectory with the minimum initial speed does not in fact touch the trunk at its topmost point. The gravitational potential energy of the grasshopper is greater at the peak of the parabola than at the uppermost point of the trunk, but its kinetic energy and total energy are smaller than they would be for a top-touching trajectory.

The numerical value of the minimal initial speed is

$$v_1^{\min} = \sqrt{2gR \left(1 + \sqrt{2} \right)} \approx 2.2 \text{ m s}^{-1}.$$

Note. (i) It is not very difficult to show that the part of the parabolic trajectory above *B* does not intersect the trunk.

(ii) We can also determine the take-off angle as $\theta = 3\pi/8 = 67.5°$, and the take-off distance as

$$AD = R(1 + \sqrt{2}/2) \approx 17 \text{ cm}.$$

(iii) It is interesting to note that point *F* is the focus of the parabola.

S39   The fleas jump in directions making angles $\frac{1}{2}(\pi - \theta)$ with the initial direction of the hair. During the period in which they are in the air, the hair, reacting to the impulsive couple it receives, rotates in the opposite direction through an angle of $\pi - \theta$, so that both fleas land on the hair but with each at the opposite end from that at which it started.

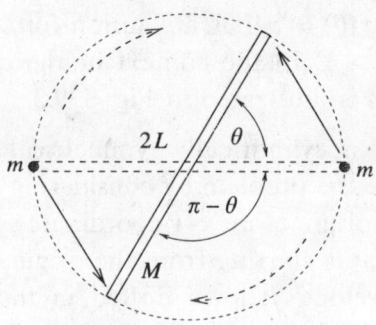

Fig. S39.1

Let v and α be the take-off speed and angle, respectively, and $2L$ be the length of the hair. The time of flight t is, as usual, $2v \sin \alpha / g$ and the range $vt \cos \alpha$. From geometry (*see* Fig. S39.1) the range must also equal $2L \sin(\theta/2)$. Now, each end of the hair receives an impulse, but only the horizontal part of the tangential component contributes to the impulsive couple acting on the hair. Thus

$$2mv \cos \alpha \cos \frac{\theta}{2} = I\omega, \quad \text{with} \quad I = \frac{1}{3}ML^2.$$

It is also necessary that $\omega t = \pi - \theta$.

Eliminating α, t and ω from the equations obtained so far, shows that θ must satisfy the equation

$$\frac{6m}{M} \sin \theta + \theta = \pi.$$

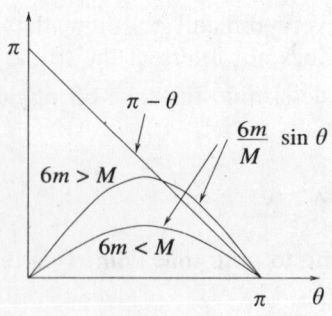

Fig. S39.2

The function $f(\theta) = n\sin\theta + \theta$ has the property $f(\pi) = \pi$, whatever the value of n. In addition, $f'(\theta) = n\cos\theta + 1 = 0$ has a solution in $0 < \theta < \pi$ provided $n > 1$. Thus, if n is strictly greater than 1, $f(\theta)$ has a maximum for some value of θ strictly less than π. This, combined with our observation about $f(\pi)$, shows that $f(\theta) = \pi$ has a solution for some value of θ strictly less than π provided $n > 1$. In the context of the question, this condition becomes $m > M/6$. This is illustrated in Fig. S39.2.

S40 The water 'bell' is cylindrically symmetrical about the vertical and so it is sufficient to solve the problem by considering a cross-section. Let the point-like rose be at the origin of an x–y coordinate system. The jets of water then follow parabolic paths starting from the origin, and our mathematical task is to find the 'envelope' (shown dotted in the figure) to this set of parabolas.

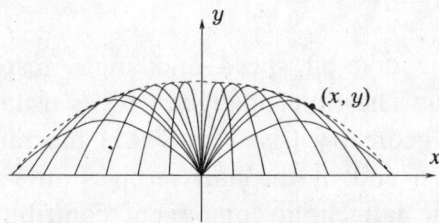

It is well known that the equation of the path of a body projected with initial speed v at an angle α to the horizontal is

$$y = x\tan\alpha - \frac{g}{2v^2\cos^2\alpha}x^2,$$

which can also be written as

$$\frac{gx^2}{2v^2}u^2 - xu + \left(y + \frac{gx^2}{2v^2}\right) = 0,$$

where $u = \tan\alpha$.

If a point (x, y) is fixed, then the above relation is a quadratic equation for u, which has a real solution if its discriminant is non-negative, i.e.

$$x^2 - 4\frac{gx^2}{2v^2}\left(y + \frac{gx^2}{2v^2}\right) \geq 0,$$

i.e.

$$y \leq \frac{v^2}{2g} - \frac{g}{2v^2}x^2.$$

This inequality divides the x–y plane into two regions separated by a parabola. Water can reach the points under the parabola (a paraboloid of revolution in three dimensions), but not those above it. The limiting parabola is the sought-for envelope.

The water 'bell' is therefore a paraboloid of revolution. It is clear from the equation for the limiting curve that the height of the 'bell' is $v^2/(2g)$, as one would expect from considering an object thrown vertically upwards. The water 'bell' defines a circle on the surface of the basin water, the radius of which can be found using the condition $y = 0$; it is $r = v^2/g$. This means that the diameter of the basin should be at least four times the height of the water 'bell' if no water is to be lost.

S41 There is uniform acceleration in both horizontal and vertical directions giving, in an obvious notation,

$$x = vt\cos\theta + \frac{1}{2}\frac{EQ}{m}t^2 \quad \text{and} \quad y = vt\sin\theta - \frac{1}{2}gt^2.$$

When $y = 0$, elimination of t results in the equation for the range given in the hint. It has a maximum value of

$$\frac{v^2}{mg^2}\left(EQ + \sqrt{m^2g^2 + E^2Q^2}\right), \quad \text{when} \quad \tan 2\theta = -\frac{mg}{EQ}.$$

The negative sign of $\tan 2\theta$ for positive Q indicates that θ needs to be more than $\pi/4$ to take advantage of the 'following wind' provided by the electric field.

Note. It will be clear that this problem is essentially equivalent to that of finding the maximum range on an inclined plane of a mass projected with a given speed.

S42 To satisfy the static equilibrium conditions for the rod (net vertical force and net torque each equal to zero) the reactions of my fingers at distances x and y from the centre of mass of the rod are (*see figure*):

$$F_x = mg\frac{y}{x + y}, \quad \text{and} \quad F_y = mg\frac{x}{x + y}.$$

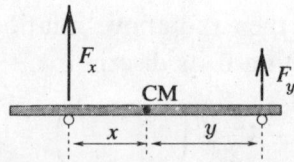

Assume that the rod first slips on my left finger. At any moment the frictional force exerted on the finger is

$$F_{\text{fr}} = \mu_{\text{kin}} F_x = \mu_{\text{kin}} mg \frac{y}{x+y}.$$

For slow movement (the horizontal acceleration is negligible) this force is equal to the static frictional force acting on the right finger, which has a maximum value of

$$\mu_{\text{stat}} F_y = \mu_{\text{stat}} mg \frac{x}{x+y}.$$

Thus the left finger can slide so long as

$$\mu_{\text{kin}} y \le \mu_{\text{stat}} x, \quad \text{i.e. } x \ge ky,$$

where $k = \mu_{\text{kin}}/\mu_{\text{stat}} \le 1$.

Initially $x_0 = y_0 = \frac{1}{2}\ell$, so my left finger slides to the position $x = x_1 = k\ell/2$ whilst working against a continuously changing frictional force. The work done during this sliding can be found using integral calculus:

$$\begin{aligned}
W(x_0 \to x_1) &= -\int_{x_0}^{x_1} \mu_{\text{kin}} F_x \, dx \\
&= -\mu_{\text{kin}} mg \int_{\ell/2}^{k\ell/2} \frac{\ell/2}{x+(\ell/2)} \, dx = mg \, \mu_{\text{kin}} \frac{\ell}{2} \ln \frac{2}{k+1}.
\end{aligned}$$

At the second stage my right finger is sliding; while $x = x_1$ and is constant, y is changing from $\ell/2$ to $y_1 = kx_1 = k^2\ell/2$.

The work done is

$$W(y_0 \to y_1) = -\int_{y_0}^{k^2 y_0} \mu_{\text{kin}} mg \frac{x_1}{x_1+y} \, dy = mg \, \mu_{\text{kin}} \frac{\ell}{2} k \ln \frac{1}{k}.$$

In the same way we can calculate the work during all of the following stages during which the rod slips alternately on my left and right finger. The total work done is

$$\begin{aligned}
W &= mg \, \mu_{\text{kin}} \frac{\ell}{2} \left[\ln \frac{2}{1+k} + (k + k^2 + k^3 + \cdots) \ln \frac{1}{k} \right] \\
&= mg \, \mu_{\text{kin}} \frac{\ell}{2} \left[\ln \frac{2}{1+k} + \frac{k}{1-k} \ln \frac{1}{k} \right].
\end{aligned}$$

If $\mu_{kin} \ll \mu_{stat}$ (i.e. $k \ll 1$), the work is done in just one step and its value is

$$W \approx \frac{\mu_{kin}\, mg}{2} \ell \times \ln 2.$$

On the other hand, if $\mu_{stat} \approx \mu_{kin}$ (i.e. $k \approx 1$), then

$$\frac{k}{1-k} \ln \frac{1}{k} \approx 1,$$

which can be confirmed either with a calculator, or by writing $k = 1 - \delta$, using $\ln(1 - \delta) \approx -\delta$, and then letting $\delta \to 0$. Thus

$$W \approx \frac{\mu_{kin}\, mg}{2} \ell.$$

S43 Take the length of the bricks to be unity and start the process from the top. The topmost brick can be displaced until half of it protrudes beyond the table, then the upper two have to be moved relative to the third one as shown in the figure. The combined centre of gravity of the upper two bricks must not be beyond the edge of the third one. Thus the second brick can only be displaced by $\frac{1}{4}$. The general strategy is to move each subpile of bricks until its combined centre of gravity is just above the edge of the brick below it.

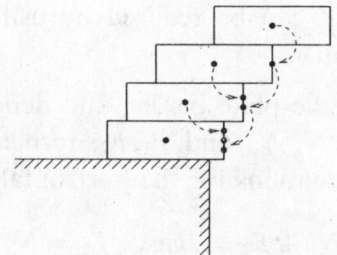

Before the third displacement, the combined centre of gravity of the top three bricks has to be found. That of the two uppermost bricks is over the edge of the third, and has to be given a double weighting, i.e. the distance of $\frac{1}{2}$ has to be divided in the ratio $2 : 1$; with the third brick being displaced by only $\frac{1}{6}$.

For the following (fourth) brick, the three placed above its edge carry triple weighting, and the distance of $\frac{1}{2}$ between the centre of gravity of the fourth brick and its edge has to be divided in the ratio $3 : 1$, i.e. the fourth brick can be displaced by only $\frac{1}{8}$. Adding up the displacements, the result is $\frac{1}{2} + \frac{1}{4} + \frac{1}{6} + \frac{1}{8} = \frac{25}{24} > 1$, in other words, the topmost brick *can* hang beyond the table.

In the above solution, critically unstable equilibrium positions have been considered at each step. In practice, the displacements should be chosen to be a little smaller than those calculated above, but it is still possible for the topmost brick to hang outside the table.

The solution can be extended to an unlimited number of bricks. The distance of $\frac{1}{2}$ between the centre of gravity and the edge of the kth brick has to be divided in the proportion $(k-1):1$, since the common centre of gravity of $(k-1)$ bricks is situated above its edge. The kth brick can therefore be displaced by a maximum of $\frac{1}{2k}$ units. If a total of n bricks is available then the displacement of the topmost one can be calculated as

$$\frac{1}{2}\left(1 + \frac{1}{2} + \frac{1}{3} + \frac{1}{4} + \cdots + \frac{1}{k} + \cdots + \frac{1}{n}\right).$$

Since the sum of the reciprocals of natural numbers tends to infinity, as is shown by considering

$$1 + \frac{1}{2} + \frac{1}{3} + \frac{1}{4} + \frac{1}{5} + \frac{1}{6} + \frac{1}{7} + \frac{1}{8} + \frac{1}{9} + \cdots$$

$$> \frac{1}{2} + \frac{1}{2} + \left(\frac{1}{4} + \frac{1}{4}\right) + \left(\frac{1}{8} + \frac{1}{8} + \frac{1}{8} + \frac{1}{8}\right) + \cdots = \frac{1}{2} + \frac{1}{2} + \frac{1}{2} + \cdots,$$

an arbitrary displacement can be realised by using a suitable number of bricks, i.e. there is no limit.

S44   Let the mass of the plate be $2m$, and denote the normal reactions and frictional forces by N_1, N_2 and F_1, F_2, respectively, as shown in Fig. S44.1. The equilibrium equations for the horizontal and vertical forces are

$$N_1 + F_2 = 2mg, \quad F_1 = N_2.$$

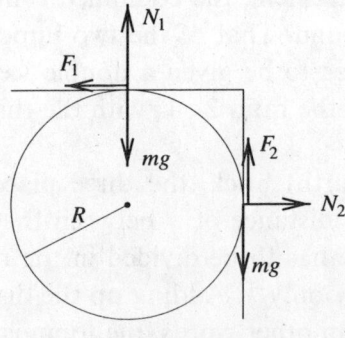

Fig. S44.1

Balancing torques about the corner of the plate gives

$$(mg + N_2)R = N_1 R.$$

The conditions controlling the frictional forces are

$$F_1 \leq \mu N_1 \quad \text{and} \quad F_2 \leq \mu N_2.$$

Using the above equations, the following three relations can be derived.

$$F_1 + F_2 = mg,$$

$$F_2 \leq \mu F_1,$$

$$F_1 \leq \mu N_1 = \mu(mg + N_2) = \mu mg = \mu F_1, \quad \text{i.e. } F_1 \leq \frac{\mu mg}{1 - \mu}.$$

In deriving the last of these relationships we have assumed that $\mu < 1$.

These three relations can be plotted in an F_1–F_2 coordinate system. If the coefficient of static friction is quite large, the situation is as shown in Fig. S44.2.

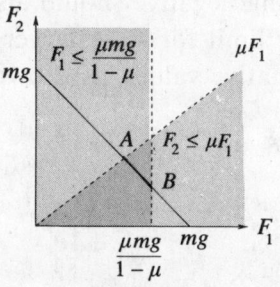

Fig. S44.2

In this case the problem does not have a unique solution; in the region represented by the straight line segment between points A and B the static equilibrium conditions can be satisfied by a range of frictional forces.

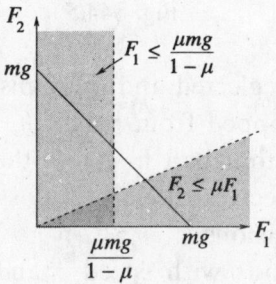

Fig. S44.3

If the coefficient of static friction is too small, the situation is as shown in Fig. S44.3 and the problem has no solution at all. The minimal possible value of the coefficient of static friction can be found by making μ such that the crossing point of the boundaries of the two inequalities occurs on the $F_1 + F_2 = mg$ straight line graph as shown in Fig. S44.4. In this case, instead of inequalities we can use equalities, and after some calculation we find the minimal value of static friction to be $\mu = \sqrt{2} - 1 \approx 0.414$.

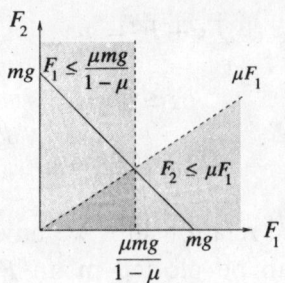

Fig. S44.4

The possibility of F_2 being negative should also be recognised. When μ is greater than 0.5 the upper limit for F_1 is larger than mg and this makes it possible for F_2 to have negative values, as shown in Fig. S44.5.

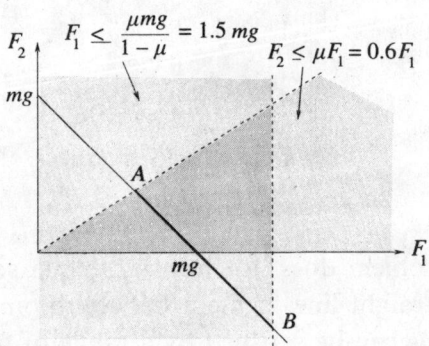

Fig. S44.5

S45 Air resistance is neglected and the balls are considered as perfectly elastic. If the balls are dropped from height h, they reach the ground with speed $v = \sqrt{2gh}$. The bottom ball first hits the ground, and then collides with the top ball, which receives the largest possible energy if the lower ball is at rest after the two collisions.

The bottom ball rebounds with speed v and collides with the top ball moving downwards at speed $-v$. Since the speed of the ball of mass m_2 is to be zero after the collision, the equations expressing the conservation of momentum and energy are

$$(m_2 - m_1)v = m_1 u \quad \text{and} \quad (m_1 + m_2)\frac{v^2}{2} = m_1 \frac{u^2}{2}.$$

The speed u of the top ball after the collision and the ratio of the masses can be calculated from these equations, giving $u = 2v$ and $m_1/m_2 = \frac{1}{3}$. With its speed doubled on rebounding, the upper ball rises to a height of $4h$.

Surprisingly, the top ball could bounce even higher than this. If $m_2 \gg m_1$ then the top ball only takes a very small fraction of the total energy after the collisions, but its speed is $3v$ and the height of the bounce is, in an ideal case, $9h$. This may sound rather incredible, but it is in agreement with the principle of conservation of energy.

Readers interested in theoretical problems may generalise the problem to n balls, while those interested in practical experimentation may try dropping sets of non-identical balls – they bounce in very amusing ways!

S46 For the first collision, momentum and energy conservation give

$$M\sqrt{2gh} = MV + \mu v,$$
$$Mgh = \frac{1}{2}MV^2 + \frac{1}{2}\mu v^2.$$

Eliminating V gives v as $2M\sqrt{2gh}/(M+\mu)$ and the kinetic energy transferred to the middle ball as $4\mu M^2 gh/(\mu+M)^2$. As a fraction of the initial energy of the first ball, this is $4\mu M/(\mu+M)^2$. The fractional energy transfer to the final ball, is the product of two such expressions using different pairs of masses. Thus in order to maximise the energy of the final ball μ should be chosen to maximise $\mu^2/(\mu+M)^2(\mu+m)^2$, i.e. $\mu = \sqrt{Mm}$. With this choice the overall fractional energy transfer is $16Mm/(\sqrt{M}+\sqrt{m})^4$ and the height attained by the final ball is $16M^2h/(\sqrt{M}+\sqrt{m})^4$.

S47 Since the dumb-bells approach each other at identical speeds, the sum of their momenta is zero in the reference frame of the air-cushioned table (the same as that of their combined centre of mass). Thus, conservation of momentum implies that the centres of mass of the two dumb-bells *always* move at identical speeds and in opposite directions.

When the dumb-bells collide, both their energy and their angular momentum are conserved, since the collision is perfectly elastic and no external torque acts on them. The states before and after the collision are shown in Fig. S47.1.

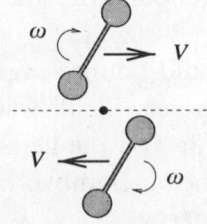

Fig. S47.1

Before the collision the dumb-bells have only translational kinetic energy, while a rotational term appears after the collision. When writing down the conservation of energy and angular momentum equations for the dumb-bells, we calculate the latter with respect to their point of contact, P:

$$2 \left(\frac{1}{2} 2mv^2 \right) = 2 \left(\frac{1}{2} 2mV^2 + \frac{1}{2} 2m\ell^2\omega^2 \right).$$

Before the collision the dumb-bells only have orbital angular momentum, but a term describing their spin about their centres of mass has to be included after the collision, i.e.

$$4\ell mv = 4\ell mV + 4m\ell^2\omega.$$

The non-trivial solution ($V \neq v$, $\omega \neq 0$) of the above set of equations is found to be $V = 0$, and $\omega = v/\ell$. That is that the centres of mass of the dumb-bells stop moving after the collision, and that the colliding point masses change velocities while the non-colliding ones keep their original velocities. This can be interpreted in the following way: point masses joined by a rigid but weightless rod are not aware of each other's presence in the course of a momentary collision. The rod only exerts a force directly after the collision, when the dumb-bell rotates about its stationary centre of mass.

The hidden point of interest in the problem is that the dumb-bells collide again after half a turn of each, i.e. after a time, $t = \pi/\omega$. Using the previous results, the resulting motion can be predicted without writing equations. The rotation of the dumb-bells stops, and they move again with the same speeds as before the first collision. Their path is the same straight line but they are now travelling 'upside down'. In other words the dumb-bells spend the time between the two collisions turning round. The speed of the dumb-bells as a function of time is shown in Fig. S47.2.

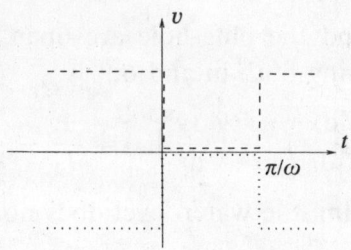

Fig. S47.2

S48 (i) At this time the two masses are travelling parallel to the x-axis in the centre of mass system and are both crossing the y-axis, one in each direction. This situation must be superimposed on that due to the motion of the centre of mass, which has moved along the line $x = L/2$ through a distance L and has a speed of $V/2$ in the y-direction. Block A is at $(\frac{1}{2}L, \frac{1}{2}L)$ with velocity $(\frac{1}{2}V, \frac{1}{2}V)$; block B is at $(\frac{1}{2}L, \frac{3}{2}L)$ with velocity $(-\frac{1}{2}V, \frac{1}{2}V)$.

(ii) Establish that the centre of mass motion is cyclic with period $8L/V$. Block A is at $(L, 50L)$ with velocity $(0, V)$; block B is at $(0, 50L)$ at rest.

S49 Let x denote the ratio of the actual water level to the level at the top of the basin; the same number shows the ratio of the current volume of water to the maximum possible volume.

During filling, x increases uniformly with time and, since it reaches the value $x = 1$ in time T_1,

$$\left(\frac{dx}{dt}\right)_{in} = \frac{1}{T_1}.$$

When water flows out, the speed of efflux – and therefore the rate of decrease of x – is proportional to the square root of the height of the column of water, i.e. to the square root of x,

$$\left(\frac{dx}{dt}\right)_{out} = -K\sqrt{x}. \tag{1}$$

The coefficient of proportionality has to be chosen so that x just decreases from 1 to 0 in time T_2.

Since the equation for the efflux is of the same form as the relation between the speed and the displacement for uniform acceleration, $v = \sqrt{2ax}$, it can be concluded that the liquid level decreases to zero at a uniformly changing speed. The initial rate of decrease is K, the final rate is zero; therefore the average rate of decrease of x equals $K/2$. This deceleration can be expressed in terms of the total time: $K/2 = 1/T_2$. The same conclusion can be reached by integrating the differential equation (1) and applying the initial and final conditions.

When both the tap and the plug-hole are open, the net rate of change caused by the water flowing both in and out is

$$\frac{dx}{dt} = \left(\frac{dx}{dt}\right)_{in} + \left(\frac{dx}{dt}\right)_{out} = \frac{1}{T_1} - \frac{2}{T_2}\sqrt{x}.$$

In a state of equilibrium, the water level does not change. The condition for this is

$$x = x_e = \left(\frac{T_2}{2T_1}\right)^2.$$

For example, if the basin fills up in the same time as it empties, $(T_1 = T_2)$ then the stationary state obtained by opening the tap and the plug-hole together corresponds to $x = \frac{1}{4}$, regardless of the initial conditions. With the data given in the problem, this ratio is $\frac{1}{9}$. It can also be seen that overflow can only be a danger if it takes more than twice as long to empty the basin, as to fill it $(T_2 > 2T_1)$.

> *Note.* One condition for the validity of Torricelli's law of efflux is that the size (diameter) of the orifice be much smaller than the depth of the water. This condition is certainly not satisfied when the basin is nearly empty, and therefore our results are only approximate. If the orifice is very small the viscosity of the water (neglected so far) also plays an important role.

S50   When the vessel is rotating the free surface of the liquid must be an equipotential surface for the system; for if it were not, the energy of the system could be lowered by changing the surface profile. The total potential energy per unit volume at any point (r, z) in cylindrical polar coordinates, is made up of two parts, the gravitational potential energy $\rho g z$ and the centrifugal potential energy. Since the centrifugal force is $\rho \omega^2 r$ directed away from the axis, the potential energy at r is the integral of this with respect to r, i.e. $-\frac{1}{2}\omega^2 r^2$. Both potentials are relative to arbitrary zeros. With the origin of z chosen as in the *Hint* and $r = 0$ taken as the zero for the centrifugal potential, the equation of the free surface is $\rho g z - \frac{1}{2}\rho\omega^2 r^2 = 0$.

If Z is the vertical distance of the lowest point of the surface below the rim of the vessel when the liquid is on the point of overflowing, then both a and Z lie on the free surface, and the volume of air in the paraboloid above the liquid but within the vessel is still one-third of the volume of the vessel.

Integrating to find this volume, $\int_0^Z \pi r^2 \, dz$ with $r^2 = 2gz/\omega^2$, gives $\pi g Z^2/\omega^2$. Thus we have

$$Z = \frac{\omega^2 a^2}{2g} \quad \text{and} \quad \frac{\pi g Z^2}{\omega^2} = \frac{\pi a^2 h}{3},$$

leading to $\Omega = (4gh/3a^2)^{1/2}$.

S51 *Solution 1.* Whilst accelerating, a car 'pushes' the Earth back a little and changes its rotational angular velocity. This very small effect has to be considered in order to resolve the paradox.

For the sake of simplicity, consider the car, of mass m, to be travelling on a body of mass M ($M \gg m$) that can move freely in the direction of the motion of the car. In the actual situation, the Earth can *rotate* freely under the car. A stationary observer would say that if the car accelerates to some speed v_0 and then subsequently to $2v_0$, the body of mass M reaches a speed $u_1 = -mv_0/M$ and then $u_2 = -2mv_0/M$, whilst its kinetic energy increases from an initial zero to $Mu_1^2/2$ and then to $Mu_2^2/2$. Since $M \gg m$, the kinetic energy of the body of larger mass and the change in its energy can be neglected. Thus the ratio of the fuel consumption values has to be $1 : 3$.

The situation is different for the observer moving with speed v_0. He can see the speed of the car increasing from v_0 to $2v_0$ and then to $3v_0$, while, in accordance with the law of conservation of momentum, the speed of the other body changes from the initial $-v_0$ to $(1 - m/M)v_0$ and then to $(1 - 2m/M)v_0$. The changes in the kinetic energy of the whole system (car plus Earth) are, therefore, firstly

$$\frac{1}{2}m\left[(2v_0)^2 - v_0^2\right] + \frac{1}{2}M\left(1 - \frac{m}{M}\right)^2 v_0^2 - \frac{1}{2}Mv_0^2 \approx \frac{1}{2}mv_0^2,$$

and, then, secondly

$$\frac{1}{2}m\left[(3v_0)^2 - (2v_0)^2\right] + \frac{1}{2}M\left(1 - \frac{2m}{M}\right)^2 v_0^2 - \frac{1}{2}M\left(1 - \frac{m}{M}\right)^2 v_0^2 \approx \frac{3}{2}mv_0^2.$$

It can be seen that the energy (fuel consumption) ratio is $1 : 3$ for both observers.

Solution 2. Friction pushes the ground backwards, i.e. it accelerates the car forward through its wheels. In the Earth's frame of reference (Peter's frame), work is done only on the car and not on the ground. On the other hand, in the train's frame of reference (Paul's frame), static friction acting on the ground also does work. Viewed from this frame, the ratio of work done on the car in the two stages is $3 : 5$, since the displacements of the car in time t are $3v_0t/2$ and $5v_0t/2$, respectively. The work done on the ground is -2

units, as the displacement of the ground is v_0t. The total work done is thus $3 - 2 = 1$ unit in the first stage of acceleration and $5 - 2 = 3$ units in the second; their ratio, when viewed from the train, is therefore again $1 : 3$.

S52 Since the object and image distances, respectively u and v, can be exchanged in the lens formula, the square of their ratio gives the magnification as $(v/u)^2 = 9$ (or $\frac{1}{9}$), implying that $v/u = 3$ (or $\frac{1}{3}$). Thus, the object distance is 30 cm (or 90 cm) and the image distance is 90 cm (or 30 cm). The focal length can be calculated from the lens formula as $f = 22.5$ cm.

If the same amount of light passed through the lens in both cases, the nine-times smaller image would be 81-times brighter, as the smaller image occupies a surface area 81-times smaller on the screen than the larger one. However, when the lens is placed at the greater distance from the source it receives only one-ninth of the light reaching it when it is close to the source. As a result, the small image is only nine-times brighter than the large one.

It can be shown in general, for such pairs of images, that the small image is always as many times brighter than the large one, as the large one is bigger in area than the small one.

S53 The lenses of the glasses of short-sited people are divergent. Let $-f$ denote the (negative) focal length of a divergent lens, d the distance between the object and the eye, and O the size of the object (*see figure*). According to the lens formula, the distance between the (virtual) image and the lens is given by

$$\frac{1}{-v} + \frac{1}{u} = \frac{1}{-f}, \qquad \text{i.e. } v = \frac{uf}{u+f},$$

whilst the size of the image is

$$I = \frac{v}{u}O = \frac{f}{u+f}O.$$

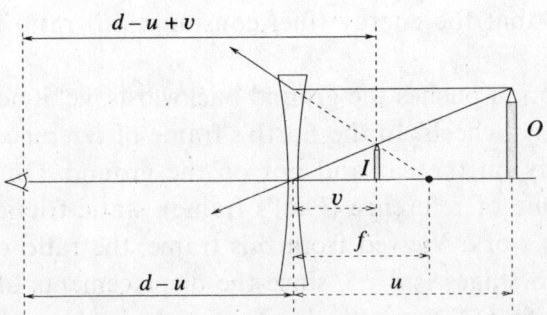

The apparent size of the image is determined by the angle ϕ it subtends at the eye, which, assuming the object is small, is

$$\phi = \frac{I}{d - u + v} = \frac{f}{u(d - u) + fd}O.$$

This angle is smallest as a function of u when the denominator on the right-hand side of the above formula is a maximum. This condition is satisfied when $u = d/2$. In other words, the apparent size of the object is smallest when the lens is equidistant from the eye and the object. It is interesting that this condition is independent of the focal length of the lens.

S54 With ϕ as defined in the *Hint*, Snell's law applied to the initial entry into the glass gives $\sin(\frac{\pi}{2} - \theta) = n_\text{g} \sin \phi$. Straightforward geometry then determines the angle between the incident ray and the normal to the glass–water interface at the point where the ray meets the boundary as $\theta + \phi$.

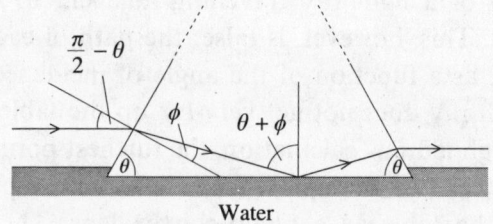
Water

For total internal reflection to occur this must exceed $\sin^{-1}(n_\text{w}/n_\text{g})$. These two conditions can be combined using the formula $\sin(\theta + \phi) = \sin \theta \cos \phi + \cos \theta \sin \phi$ to eliminate ϕ and obtain

$$n_\text{g}^2 - n_\text{w}^2 \geq \cos^2 \theta(n_\text{g}^2 + 1 - 2n_\text{w}).$$

Substitution of the given values for the refractive indices yields the stated result.

S55 Consider the light beam as consisting of parallel light rays. They cross the vertical plane face of the quarter-cylinder without changing their direction, and strike the curved surface of the cylinder at various angles of incidence. The normals at the points of incidence of the rays are radii of the cylinder.

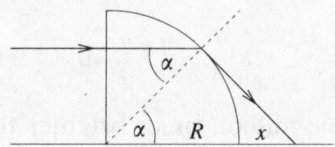

The higher the position of a light ray entering the quarter-cylinder, the larger is its angle of incidence at the cylinder's curved surface. The angle of incidence for the ray shown in the figure is the critical angle for total internal reflection. Therefore only light rays closer to the table than this one can leave the quarter-cylinder (refracted to different extents). The limiting case is determined using the figure:

$$\sin \alpha_h = \frac{1}{n} = \frac{2}{3} \quad \text{and} \quad \frac{R}{R+x} = \cos \alpha_h,$$

which yield $x = 1.71$ cm. This is the closest to the quarter-cylinder that light can reach the table.

As the angle of incidence of light rays close to the table top is smaller, they are deviated less from their original direction by refraction, and therefore might reach the surface of the table further away. One is inclined to think that, in principle, the light patch could reach to any distance along the table, since the direction of a light ray travelling adjacent to the surface of the table is not altered. This, however, is false; the path of each light ray can be parameterised (e.g. as a function of the angle of incidence), and it can then be shown that each ray does not get very far up the table.

Instead of through tedious calculation, the furthest point of the light patch can be found by means of a simple 'trick'. Consider the part of the quarter-cylinder close to the table as a plano-convex lens. The cylinder material before the lens behaves like a plano-parallel plate and can be ignored. The focal length of the plano-convex lens can be calculated using the thin lens formula:

$$\frac{1}{f} = \frac{n-1}{R}.$$

This yields $f = 10$ cm, and this is the distance from the quarter-cylinder of the furthest point of the light patch.

S56 If R_M is the Moon's radius and R is the Moon–Earth distance, the light power diffusely reflected into a solid angle of 2π is $\alpha \pi R_M^2 E$, where E is the intensity of direct sunlight (on either the Earth or the Moon). The intensity received on Earth as moonlight is this divided by $2\pi R^2$. The Moon's diameter subtends about $\frac{1}{2}^\circ$, or 9×10^{-3} rad, at the Earth's surface, and so the ratio of moonlight intensity to that of direct sunlight is

$$\frac{\alpha R_M^2}{2R^2} = \frac{0.07}{2} \times \frac{1}{4} \times \left(9 \times 10^{-3}\right)^2 \approx 10^{-6}.$$

Thus sunlight is about one-million-times brighter than moonlight.

Note. In fact, the reflectivity of the Moon was actually measured by comparing the brightnesses of sunlight and moonlight. Moreover, the albedo of the Earth could be similarly determined by measuring the (very low) brightness of the dark part of a new moon illuminated by reflected light from the Earth.

S57 Let us approximate comfortable walking by a model in which the human leg is a freely swinging pendulum. The period of a freely swinging body supported at its upper end is

$$T = 2\pi\sqrt{\frac{I}{mgs}},$$

where I is the moment of inertia of the body, m is its mass, and s is the distance between the pivot and the centre of mass of the body. Now introduce the so-called effective length $L_{\text{eff}} = I/ms$ by expressing the period as

$$T = 2\pi\sqrt{\frac{L_{\text{eff}}}{g}}.$$

We can assume that the effective length is directly proportional to the actual length of the leg and, for different people with the same leg length, we will find only very slight differences in the effective length.

Using this result for the natural period, we can now estimate a person's natural gait – the one involving the least muscular effort. To a first approximation, we assume that the length of a stride is proportional to the length of the leg. The time for a single stride is one-half the period given above, and the walking speed v depends on the leg length:

$$v_{\text{walk}} \propto \frac{L}{T/2} \propto \sqrt{L}.$$

This equation predicts that people with longer legs have a more rapid natural walking gait. The prediction is made on the basis of an oversimplified model that assumes minimum energy expenditure and ignores differences in anything (e.g. shape, strength, etc.) other than leg length. However, the prediction is borne out by common experience.

When analysing a person running, an important change must be made to our model. During running, the leg does not swing freely but is subjected to a torque acting about its pivot. The torque is produced by a force F supplied by the muscles. This force is roughly proportional to the cross-sectional area of the muscles involved, and if we assume that, for people of different size, the relative proportions of the leg are the same, then the cross-sectional area, and therefore the force F, depends on the square of the length L. The torque is then proportional to the product of F and L:

$$\tau \propto FL \propto L^2 \times L = L^3.$$

The moment of inertia I is proportional to the mass and to the square of the length. Again, we assume that all legs have essentially the same proportions; that is, width and thickness are proportional to length. Thus the mass varies as the cube of the length and

$$I \propto mL^2 \propto L^5.$$

It can generally be shown that for a body oscillating about a fixed point and subject to a periodic torque, the period T depends on the maximum torque τ and the moment of inertia I of the body about that point, and is given by

$$T \propto \sqrt{\frac{I}{\tau}}.$$

Upon substituting for I and τ, we find

$$T \propto \sqrt{\frac{L^5}{L^3}} \propto L.$$

The speed of running is the product of the frequency of taking steps and the length of a single step, and hence

$$v_{\text{run}} \propto f \times L \propto \frac{L}{T} \propto \frac{L}{L} = 1.$$

So the model predicts, in accord with Annie and Andy's experience, that the speed of running does *not* depend on leg length. Whilst its predictions are not, of course, strictly accurate, the model does offer some explanation for the observation that the ordinary walking rate of people with long legs is usually greater than that of people with short ones, whereas the speed at which they can run is often not significantly different.

S58   Consider a simple pendulum of length L and a pendulum consisting of a uniform rod of length ℓ pivoted at one end. If both are released from a horizontal position, what are their angular speeds after they have each travelled through an angle α?

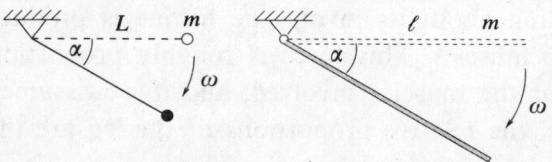

The principle of conservation of energy yields

$$\frac{1}{2}mL^2\omega^2 = mgL\sin\alpha, \quad \text{i.e.} \quad \omega = \sqrt{\frac{2g}{L}\sin\alpha}$$

for the simple pendulum, and

$$\frac{1}{2}\frac{m\ell^2}{3}\omega^2 = mg\frac{\ell}{2}\sin\alpha, \qquad \text{i.e.} \quad \omega = \sqrt{\frac{3g}{\ell}\sin\alpha}$$

for the rod. If $L = \frac{2}{3}\ell$, then the angular velocities of the two motions are equal for all values of α. It then follows that the two motions are identical at all times and their periods are equal.

How can the period of this equivalent pendulum be calculated? The formula $T = 2\pi\sqrt{L/g}$, valid for small oscillations, cannot be applied as the amplitude here is large. Exact calculations would require complicated mathematical analysis, but this is not necessary if, instead of calculating the period T, we only wish to determine its dependence on L.

The period of swing of the simple pendulum may depend on its length L, the mass of its bob m, the gravitational acceleration g and the maximum angle of deviation α_{max}. If the dimensions of the quantities involved are taken into consideration, this functional dependence can only be of the form

$$T(L, m, g, \alpha_{max}) = f(\alpha_{max})\sqrt{\frac{L}{g}}.$$

To justify this assertion, we note the following points. The dimension of mass is the 'kilogram', and since the 'kilogram' does not occur in the dimensions of any of the other quantities, the period (which has dimension 'seconds') cannot depend upon the mass of the bob. On the other hand, 'seconds' occur only in g, and therefore the required dimension of 'seconds' in T can only be obtained if T is inversely proportional to the square root of g. Finally, in order to settle the 'metre' dimension, the period has to be proportional to the square root of L. The form of the function $f(\alpha_{max})$ cannot be determined via dimensional analysis, since the angle is dimensionless. The only available information is that for small angles $f(\alpha_{max}) \approx 2\pi$.

From the above reasoning, it can be concluded that (with the same initial displacements) the period of a simple pendulum of length $(2/3)\ell$ is $\sqrt{2/3}$ times that of a simple pendulum of length ℓ. Thus, the period of a pivoted rod of length ℓ is approximately 82 per cent of that of a simple pendulum of the same length. This conclusion is valid not only for horizontal release, but for any common initial starting position.

S59 The required power for the hovering helicopter depends on the gravitational acceleration g, the linear size of the helicopter L, the average density of the helicopter ρ_{hel}, and the density of air ρ_{air}.

It is reasonable to assume that the mechanical power needed depends only on these quantities and that the dependence is a power relationship:

$$P \propto g^\alpha \times L^\beta \times \rho_{hel}^\gamma \times \rho_{air}^\delta.$$

The dimensions of the left- and right-hand sides must be equal:

$$\frac{\text{kg m}^2}{\text{s}^3} = \left(\frac{\text{m}}{\text{s}^2}\right)^\alpha \times \text{m}^\beta \times \left(\frac{\text{kg}}{\text{m}^3}\right)^\gamma \times \left(\frac{\text{kg}}{\text{m}^3}\right)^\delta,$$

which yields

$$\gamma + \delta = 1,$$
$$\alpha + \beta - 3(\gamma + \delta) = 2,$$
$$-2\alpha = -3.$$

The solution of this system of linear equations is $\beta = \frac{7}{2}$, $\alpha = \frac{3}{2}$ and $\gamma = 1 - \delta$.

It can be seen that the mechanical power needed is proportional to the $\frac{7}{2}$ power of the linear size. Consequently, the second helicopter should have an engine producing power $(1/2)^{7/2} P = 0.088P$.

> *Note.* (i) The efficiency of a mechanical engine can be characterised by the ratio of the power produced to the mass of the engine. According to the above result the 'specific power'
>
> $$\frac{P}{m} \propto \frac{P}{L^3} \propto \sqrt{L},$$
>
> i.e. the efficiency required increases as the linear size increases. This means that the smaller a helicopter is, the more easily it can hover. There are many small animals (bees, dragonflies, hummingbirds, etc.) that can hover like a helicopter, but larger birds are unable to do so.
>
> (ii) Using simple dimensional analysis we could find only the *sum* of the exponents γ and δ. However, it is clear that P can depend only on the product of the density of the helicopter and g, because, when the helicopter is hovering, the relevant quantity is not its inertial mass, but its weight. Thus γ must be equal to α, i.e. $\gamma = \frac{3}{2}$ with $\delta = -\frac{1}{2}$. Finally, we get
>
> $$P \propto (g\rho_{hel})^{3/2} \times L^{7/2} \times \rho_{air}^{-1/2} = (L^3 \rho_{hel} g) \times \sqrt{Lg} \times \sqrt{\frac{\rho_{hel}}{\rho_{air}}}.$$

Here, on the surface of the Earth, we can change only the size and density of the helicopter. Nevertheless, for a space mission using robot helicopters, it could be useful to know how P depends on the gravitational acceleration and the atmospheric density of the target planet.

S60 The gravitational potential energy lost as the rod falls through an angle θ is $Mg\frac{\ell}{2}(1-\cos\theta)$, and this is converted into rotational kinetic energy about the edge of the table. Either by direct calculation or by using the parallel axes theorem, the relevant moment of inertia of the rod is found to be $\frac{1}{3}M\ell^2$. Combining these two results gives

$$\omega^2 = \frac{3g}{\ell}(1 - \cos\theta).$$

The centripetal acceleration a_c, is $\ell/2$ times ω^2 and therefore equal to $\frac{3}{2}g(1-\cos\theta)$. Using the same moment of inertia and the instantaneous torque of $Mg\frac{\ell}{2}\sin\theta$ gives the tangential acceleration of the rod as $a_t = \frac{3}{4}g\sin\theta$.

(i) The smooth (frictionless) horizontal and vertical walls of the groove can exert only positive vertical and horizontal forces, V and H respectively, on the end of the rod, (*see* Fig. S60.1). The rod will lose contact with the table as soon as one of these falls to zero and is required by the equations of motion to become negative, i.e. to become a (physically impossible) force of attraction.

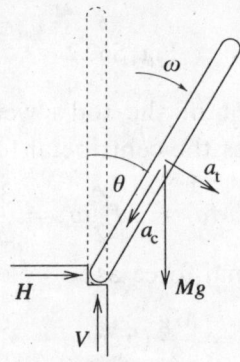

Fig. S60.1

Resolving forces and accelerations horizontally and vertically (Fig. S60.1) gives

$$H = M(a_t\cos\theta - a_c\sin\theta),$$
$$Mg - V = M(a_c\cos\theta + a_t\sin\theta).$$

Solving these two equations for H and V we obtain

$$H = \frac{3}{4}Mg\sin\theta\,(3\cos\theta - 2),$$

$$V = \frac{1}{4}Mg\,(3\cos\theta - 1)^2.$$

The first reaction component to vanish is H and this happens when $\theta = \cos^{-1}\frac{2}{3} \approx 48°$. At larger angles, H would be negative, so the rod really loses contact with the table because the smooth groove is not able to pull back the end of the rod.

(ii) As the edge of the table is now a very small quarter-circle, the normal force N is always directed along the rod's axis. The static frictional force F_{fr} is tangential to this quarter-circle and can have any arbitrary value because of the rough edge (*see* Fig. S60.2).

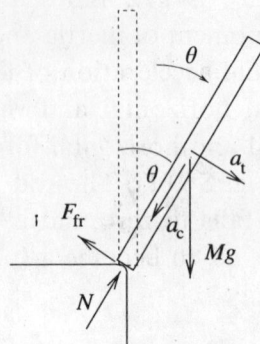

Fig. S60.2

The sum of the component of the rod's weight along the rod and the normal force of the table gives the centripetal force:

$$Mg\cos\theta - N = Ma_c = M\frac{\ell}{2}\omega^2 = \frac{3}{2}Mg(1 - \cos\theta).$$

We can thus express the normal force as

$$N = \frac{Mg}{2}(5\cos\theta - 3).$$

The reaction of the table on the rod becomes zero when $N = 0$, i.e. when $\theta = \cos^{-1}\frac{3}{5} \approx 53°$. At larger angles the normal force should be negative, which is impossible, and thus the rod loses contact with the table. Because of the rough edge, the static frictional force is always large enough to prevent slipping except when the normal force becomes zero; consequently, it has no effect on the motion.

> *Note.* In this problem we considered two extreme cases represented by (i) and (ii). In general, the direction of the normal force is perpendicular to the common tangent to the table's edge and the bottom of the rod; this means that the normal force can act in virtually any direction. This shows that the motion of the falling rod strongly depends on the geometrical details of the touching surfaces, as well as on the value of the coefficient of friction (*see also* P61).

S61 Suppose, first, that the surface of the table is very smooth (the coefficient of friction is very small). Just after the pencil is released, its centre of mass accelerates in the direction of the fall and acquires both horizontal and vertical velocities. The horizontal component of the acceleration is produced by the frictional force between the pencil point and the table, but, since the surface of the table is smooth, the point soon slips, in the direction *opposite to that of the fall*.

If the friction is very large, the pencil does not slip for a relatively long time. Initially, the horizontal velocity of the centre of mass, which is moving on a circular path, increases in the direction of the fall, but later it starts to decrease and, if the pencil continued moving this way until it was horizontal, it would tend to zero. The sign of the horizontal acceleration of the centre of mass, and thus also that of the frictional force, changes during the motion. If the pencil does not slip during the first stage, then it can only slip 'forward', i.e. *in the direction of the fall*.

In the following we are going to prove that the pencil will slip in some way ('backward' or 'forward') but the point of the pencil never loses contact with the table. For sake of mathematical simplicity let us use quantities with a value of unity for the length and mass of the pencil, as well as for the gravitational acceleration: $\ell = M = g = 1$. Thus the weight of the pencil is 1, its centre of mass (CM) is $\frac{1}{2}$ measured from either end, its moment of inertia about its CM is $\frac{1}{12}$, and its moment of inertia about one of its ends is $\frac{1}{3}$.

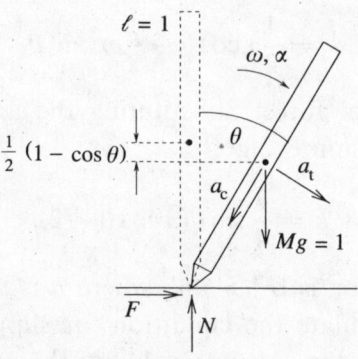

Fig. S61.1

During the first stage of the motion the point of the pencil does not slip, and so the pencil rotates about its point (Fig. S61.1). We can find the angular velocity of the pencil with the help of the conservation of energy:

$$\frac{1}{2}(1 - \cos\theta) = \frac{1}{2} \times \frac{1}{3}\omega^2,$$

which yields $\omega = \sqrt{3(1 - \cos\theta)}$. The instantaneous torque of $\frac{1}{2}\sin\theta$ gives the angular acceleration of the pencil:

$$\frac{1}{2}\sin\theta = \frac{1}{3}\alpha, \qquad \text{i.e. } \alpha = \frac{3}{2}\sin\theta.$$

Vertically there are two forces acting on the pencil: its weight and the normal force of the table, N. The vertical component of the centripetal acceleration of CM is $\frac{1}{2}\omega^2\cos\theta$ and that of the tangential acceleration is $\frac{1}{2}\alpha\sin\theta$. Thus the vertical component of the equation of motion is

$$1 - N = \frac{1}{2}\alpha\sin\theta + \frac{1}{2}\omega^2\cos\theta,$$

which yields:

$$N = \left(\frac{3\cos\theta - 1}{2}\right)^2.$$

It seems that the normal force is never negative, and that the point of the pencil cannot lose contact with the table during the rotational phase of the motion. The normal force is zero when $\theta = \cos^{-1}\frac{1}{3} \approx 70.5°$, which means that the frictional force is also zero at this angle, and the pencil will slip there if it has not done so before.

The horizontal component of the equation of motion is

$$F = \frac{1}{2}\alpha\cos\theta - \frac{1}{2}\omega^2\sin\theta,$$

where F is the frictional force. Substituting the angular acceleration and velocity into this expression, we get

$$F = \frac{3}{4}\sin\theta(3\cos\theta - 2).$$

The condition for slipping is $|F| > \mu N$, where μ is the coefficient of (static) friction. We can reformulate the condition for slipping with the help of a function $f(\theta)$, defined as the absolute value of the ratio of the forces F/N,

$$f(\theta) = \left|\frac{F}{N}\right| = \left|\frac{3\sin\theta(3\cos\theta - 2)}{(3\cos\theta - 1)^2}\right| > \mu.$$

The function $f(\theta)$ is plotted in Fig. S61.2.

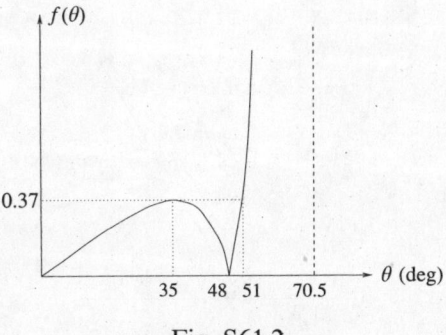

Fig. S61.2

The frictional force changes sign at $\theta = \cos^{-1} \frac{2}{3} \approx 48°$, implying that 'backward' slipping can occur in the region $0 < \theta < 48°$. Using numerical methods it can be shown that in this region $f(\theta)$ has a maximum value of $\mu_{crit} \approx 0.37$ at an angle of $\theta \approx 35°$. It means that the pencil slips 'backward', if $\mu < \mu_{crit}$.

If $\mu > \mu_{crit}$ then the pencil slips 'forward' before it reaches the angle of $\theta \approx 70.5°$, where $f(\theta)$ approaches infinity. (Note that the pencil cannot start slipping in the range of $35° < \theta < 51°$.) The 'backward' and 'forward' motions of the pencil are shown in Fig. S61.3. In both cases the slipping point of the pencil can stop again.

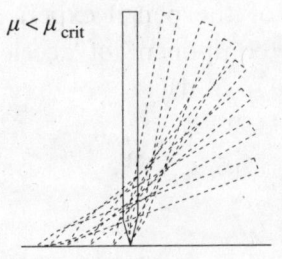
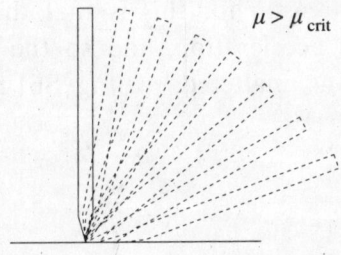

$\mu < \mu_{crit}$

$\mu > \mu_{crit}$

Fig. S61.3

Finally, it will be shown that the point of the pencil does not lose contact with the table. Let us consider first the case of 'forward' slipping shown in Fig. S61.4. According to the work–energy theorem and the cosine law for (vector) triangles,

$$\frac{1}{2}(1 - \cos\theta) = W_{fr} + \frac{1}{2} \times \frac{1}{12}\omega^2 + \frac{1}{2}\left[\left(\frac{\omega}{2}\right)^2 + v_{point}^2 + 2v_{point} \times \frac{1}{2}\omega\cos\theta\right].$$

If we neglect the work of friction and the two terms containing the velocity, v_{point} of the point of the pencil, (all three terms are positive), then we obtain

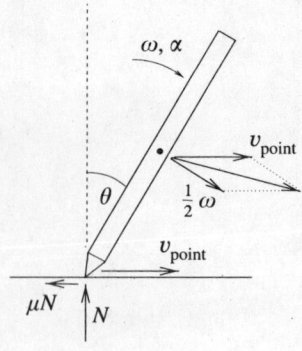

Fig. S61.4

an inequality for the angular velocity: $\omega^2 < 3(1 - \cos\theta)$. The instantaneous torque about CM gives

$$\frac{1}{2}N(\sin\theta + \mu\cos\theta) = \frac{1}{12}\alpha,$$

whilst the vertical component of the equation of motion is

$$1 - N = \frac{1}{2}\alpha\sin\theta + \frac{1}{2}\omega^2\cos\theta.$$

(It will be recognised that this equation is the same as for the non-slipping case. The reason for this is that the point of the pencil experiences only horizontal acceleration, and so the vertical component of acceleration of CM remains unaltered, *see* Fig. S61.5.)

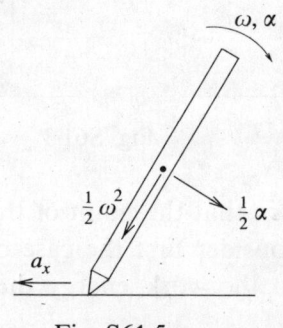

Fig. S61.5

From the two equations above we can express the normal force as a function of θ and ω^2. However, we have an inequality for ω^2 which yields

$$N = \frac{1 - (1/2)\omega^2\cos\theta}{1 + 3\sin\theta(\sin\theta + \mu\cos\theta)} > \frac{(3/2)[\cos\theta - (1/2)]^2 + (5/8)}{1 + 3\sin\theta(\sin\theta + \mu\cos\theta)} > 0.$$

Thus the normal force is always positive, and the point of the pencil does not lose contact with the table.

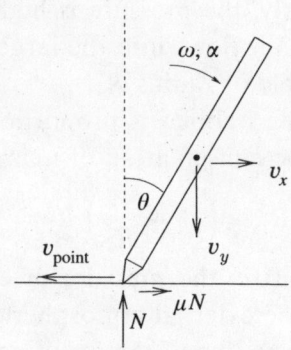

Fig. S61.6

For the case of 'backward' slipping our method is very similar. Figure S61.6 shows the horizontal and vertical components of the velocity of CM, v_x and v_y, respectively. As the point of the pencil has zero vertical velocity, there is a connection between v_y and ω, namely $v_y - \frac{1}{2}\omega \sin \theta = 0$. We can again use the work–energy theorem to give

$$\frac{1}{2}(1 - \cos \theta) = W_{\text{fr}} + \frac{1}{2}\left(\frac{1}{12}\omega^2 + v_x^2 + v_y^2\right),$$

in which we (again) neglect work against friction and another positive term containing v_x to obtain

$$\omega^2 < \frac{1 - \cos \theta}{(1/12) + (1/4)\sin^2 \theta}.$$

Considering again the net torque about CM and the vertical component of the equation of motion, we obtain a further inequality for the normal force, namely:

$$N = \frac{1 - (1/2)\omega^2 \cos \theta}{1 + 3\sin \theta(\sin \theta - \mu \cos \theta)} > \frac{1 + 3(\cos \theta - 1)^2}{[1 + 3\sin \theta(\sin \theta - \mu \cos \theta)](1 + 3\sin^2 \theta)}.$$

The numerator is always positive and the denominator is positive (for $0 \le \theta \le 90°$), if $\mu < \frac{4}{3}$. However, in the 'backward' slipping region $\mu < 0.37$, so, again, the point of the pencil does not lose contact with the table.

If the pencil point stops slipping at some stage, it cannot lose contact again, because then $\omega^2 < 3(1 - \cos \theta)$, and thus $N > (3\cos \theta - 1)^2/4 \ge 0$ (see the first part of this solution).

S62   The pressure in a soap bubble of radius R is greater than the atmospheric pressure p_0 by $\Delta p = 4\gamma/R$. The factor of 4 arises because both directions of curvature and the fact that the film is two-sided have to be taken into account. Clearly, the pressure is higher in the bubble of smaller radius and, therefore, the air flows into the larger bubble, as if inflating it to finally form a single bubble of radius R_3.

The volume of air in the bubbles is proportional to R^3, and therefore the ideal gas equation and the conservation of mass require that

$$\left(p_0 + \frac{4\gamma}{R_1}\right) R_1^3 + \left(p_0 + \frac{4\gamma}{R_2}\right) R_2^3 = \left(p_0 + \frac{4\gamma}{R_3}\right) R_3^3.$$

For bubbles of ordinary size, the pressure of curvature is many orders of magnitude smaller than the external atmospheric pressure. If the pressure of curvature is neglected the radius of the resulting bubble is

$$R_3 \approx \sqrt[3]{R_1^3 + R_2^3}.$$

If the radii are measured 'accurately', in order to determine the surface tension, the formula

$$\gamma = \frac{p_0}{4} \frac{R_3^3 - R_1^3 - R_2^3}{R_1^2 + R_2^2 - R_3^2}$$

is appropriate. In practice, however, this method cannot be applied, as the numerator is, as shown, almost equal to zero and thus would carry a large fractional uncertainty as a result of measurement error. Any measured data are likely to provide only a rough estimate of the surface tension.

S63   The cross-sectional edge of the disc of water is a semicircle of radius $r = \frac{1}{2}d$ (*see figure*). Thus, the curvature of the surface of the water is $2/d$, which corresponds to a pressure of curvature of $\Delta p = 2\gamma/d$, where γ is the surface tension. (The other component of the curvature is negligible because $D \gg d$.)

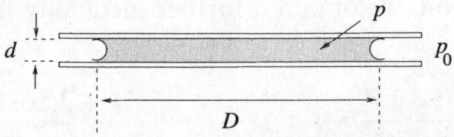

The pressure inside the disc is therefore $p_0 - 2\gamma/d$ when the atmospheric pressure is p_0. This pressure difference acts over a surface area between the water and each of the glass surfaces of $\pi D^2/4$. This implies that a force,

$$F = \frac{\pi D^2}{4} \frac{2\gamma}{d}$$

'pulls' the glass plates together.

Note. If d is much smaller than D, this force can be quite considerable. It is in fact very difficult (if not impossible) to separate two parallel glass plates by pulling them in a direction perpendicular to their common plane when there is water between them. In order to be separated, they have to be slid in a direction parallel to that plane.

S64 The velocity of the thread at a distance of x metres from the wall is obviously proportionately smaller than the velocity of the end of the thread, i.e. it is xv_0.

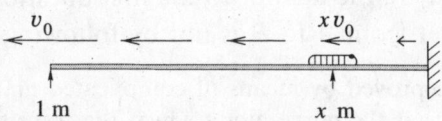

If this value is greater than the speed of the caterpillar, then the latter will move away from the wall. Its situation will become more and more hopeless, and it will never reach the wall.

On the other hand, if $v_{\text{caterpillar}} > xv_0$, the net velocity of the caterpillar is towards the wall and increases as time passes, with the consequence that the caterpillar will certainly reach the wall. The limiting case corresponds to $x = v_{\text{caterpillar}}/v_0 = 0.1$ m. Starting at this point, the caterpillar does not move in either direction.

S65 Imagine signs attached to points on the thread and labelled with the ratio of the distance from the wall to the total length of the thread. These figures are precisely the coordinates x mentioned in the solution to the previous problem: $x = 0$ corresponding to the position of the wall and $x = 1$ to that of the spider. Now, however, these 'stretch' as the thread stretches.

We first calculate how long it takes the caterpillar to get from a point x to a nearby point $x - \Delta x$ when the spider has been moving for a time t. Since the distance between the points in question is $(1 + v_0 t)\Delta x$ and the caterpillar moves at speed c, the relationship

$$\Delta x = \frac{c\Delta t}{1 + v_0 t}$$

holds. Summing (integrating) this relationship for the whole motion of the caterpillar, which starts from x_0 and reaches the wall in time T, gives

$$x_0 = \sum \frac{c\Delta t}{1 + v_0 t} \approx \int_0^T \frac{c}{1 + v_0 t}\, \mathrm{d}t = \frac{c}{v_0} \ln(1 + v_0 T).$$

Since the above integral can be made arbitrarily large by a suitable choice of T, the perhaps surprising result is that however quickly the spider pulls the end of the thread (i.e. with an arbitrarily large v_0, an arbitrarily small c and an arbitrary x_0), the caterpillar will still reach the wall within a finite time.

S66   (i) Suppose that the ball falls freely for 1 m, then reaches point B by bouncing practically horizontally along a row of closely spaced nails near the bottom edge of the drawing-board. The duration of the vertical fall is $t_1 = 0.45$ s, at the end of which the ball has reached a speed of $v_1 = 4.4$ m s^{-1}, and covers the remaining distance of 2 m in a time $t_2 = 0.45$ s. As the ball would have reached point B in a time $t_3 = 1.01$ s by sliding down the straight line AB (with acceleration $g/\sqrt{5}$), the answer to the first question is that the quickest way for the ball to get from A to B is not by following the shortest route.

> *Note.* It can be proved by means of complicated mathematics (the calculus of variations) that the curve along which the transit time is the shortest is a cycloid.

(ii) A body dropped from rest at point A has maximum vertical velocity if it is in free fall. Its maximum kinetic energy, and therefore its maximum speed, is determined solely by the magnitude of its vertical displacement. Thus, a bouncing ball cannot reach the bottom of the drawing-board faster than a body in free fall, i.e. in less than $t_1 = 0.45$ s. The answer is therefore *no*!

S67   The puzzling aspect of the problem is that insufficient data have been given in the text. However, the figure can be used as a source of information. Using a protractor you can measure with sufficient accuracy that the tangent to the fixed end of the rope makes an angle of 30° with the vertical, as shown in the figure. This means that the tension at the fixed end of the rope is $T = 20$ N$/\sin 30° = 40$ N. Similarly, the weight of the rope is equal to the vertical component of the tension there; $mg = T\cos 30° = 34.6$ N, giving the mass of the rope as $m = 3.5$ kg.

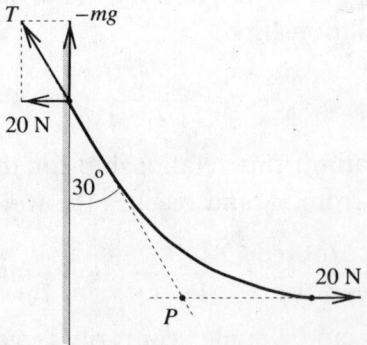

Note. Further information can be obtained from the figure. The centre of gravity of the rope must be vertically above the point P, because the lines of action of all three forces acting on the rope must meet at a single point.

S68 Using the argument described in the *Hint* it is possible to prove that the compasses have to be opened to the extent necessary to make the lower arm hang horizontally when the compasses are suspended, as shown in Fig. S68.1.

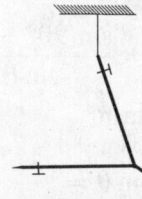

Fig. S68.1

Starting from that situation, let us imagine for the moment that the upper arm of the compasses is fixed. If the lower arm is then bent *either* upwards *or* downwards, the horizontal position of the CM of the compasses moves towards the pivot. After the release of the upper arm the pivot moves downwards, because that is the only way that the CM of the compasses can again position itself below the attachment point. So, in either case, the vertical position of the pivot is lowered, and we can conclude that the originally described situation is the one required.

Instead of a real pair of compasses let us consider a simplified model of two identical thin rods joined by a pivot of negligible mass as shown in Fig. S68.2.

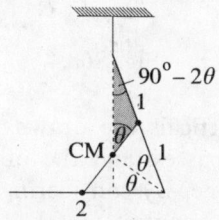

Fig. S68.2

Let the angle between the arms of the compasses be 2θ, and the length of each of the arms be 2 units. It is easy to find congruent angles in Fig. S68.2 and to apply the sine rule to the shaded triangle. Figure S68.3 shows a magnified version of this triangle.

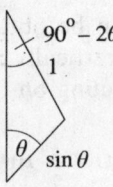

Fig. S68.3

According to the sine rule it follows that

$$\frac{\sin\theta}{1} = \frac{\sin(90° - 2\theta)}{\sin\theta}.$$

After a simple calculation we obtain

$$\sin\theta = \frac{1}{\sqrt{3}},$$

so $\theta \approx 35.3°$, that is to say the compasses should be opened to $2\theta \approx 70.5°$.

Note. The result $2\theta = 2\sin^{-1}\left(1/\sqrt{3}\right) = \cos^{-1}(1/3)$ is so simple, one might suppose that a more elegant solution exists. Actually, it is possible to find the angle 2θ with the help of the theorem of parallel transversals.

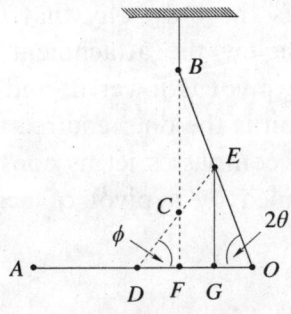

Fig. S68.4

In Fig. S68.4 two verticals are drawn, one through the overall centre of mass C of the compasses, and the other through the centre of mass E of the upper arm alone. By considering the lines forming the angle 2θ, it can be seen that the equality of OE and EB implies that $FG = GO$. Similar consideration of the angle ϕ, and the fact that $DC = CE$, shows that $FG = DF$. Thus points F and G trisect the section OD, which in turn is equal to OE, thus implying that $OE = 3GO$. As EGO is a right-angled triangle, it follows that $2\theta = \cos^{-1}(1/3) \approx 70.5°$.

S69   Clearly, the centre of mass S of the triangle has to be below the point of suspension. Denote the vectors pointing from the centre of mass

S to the vertices of the triangle by $\mathbf{r}_1, \mathbf{r}_2$ and $\mathbf{r}_3$, and that to the suspension point by $\mathbf{m}$ (*see figure*). The forces $\mathbf{F}_1, \mathbf{F}_2$ and $\mathbf{F}_3$ exerted on the plate by the threads can now be expressed in terms of the vectors defining the thread directions:

$$\mathbf{F}_i = \lambda_i(\mathbf{m} - \mathbf{r}_i), \quad i = 1, 2, 3.$$

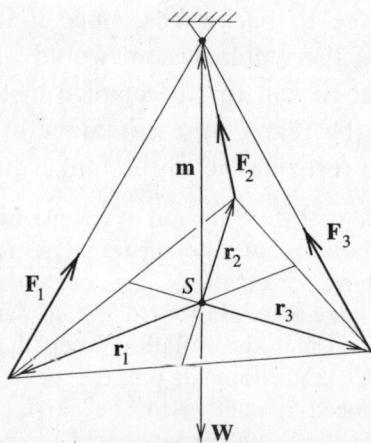

Since the plate is in equilibrium, the vector sum of the forces acting on it is zero, i.e.

$$\mathbf{F}_1 + \mathbf{F}_2 + \mathbf{F}_3 + \mathbf{W} = \mathbf{0}.$$

Making use of the fact that the vector pointing to the centre of mass of the triangle (the origin of the vector reference frame) is the arithmetic mean of the vectors pointing to its vertices, we have

$$\mathbf{r}_1 + \mathbf{r}_2 + \mathbf{r}_3 = \mathbf{0}.$$

We note that $\mathbf{W}$ and $\mathbf{m}$ are parallel and, therefore, $\mathbf{W} = -k\mathbf{m}$. Eliminating $\mathbf{r}_3$ from the above equations gives

$$(\lambda_3 - \lambda_1)\mathbf{r}_1 + (\lambda_3 - \lambda_2)\mathbf{r}_2 + (\lambda_1 + \lambda_2 + \lambda_3 - k)\mathbf{m} = \mathbf{0}.$$

Since $\mathbf{r}_1$, $\mathbf{r}_2$ and $\mathbf{m}$ are not in the same plane, a linear combination of them can only be zero if the coefficient of each is zero.

Thus $\lambda_1 = \lambda_2 = \lambda_3$, which implies that the tensions in the threads are proportional to their lengths. This deduction would become invalid if one of the threads were slack, since the plane of the plate would then become vertical and m would lie in it.

S70 Initially the tanker and the liquid in it are at rest. As the outlet pipe is at the rear of the tanker, when the tap is opened the centre of mass of the liquid will move backwards. As the centre of mass of the whole system is fixed, the tanker itself must move forward. However, the outlet pipe is vertical and so the emerging liquid will acquire a forward horizontal-velocity component. This does not contradict the law of conservation of linear momentum, because the liquid inside the tanker will be moving backwards (relative to the ground). Nevertheless, the forward direction of motion of the tanker must subsequently change to backwards, since if it did not, the position of the centre of mass of the whole system would ultimately start to move forwards. The dynamical reason for the change in the direction of motion is the force exerted on the rear of the container by the backward-moving liquid as it is brought to rest relative to the tanker just before discharge.

> *Note.* (i) It could be that the direction of travel of the tanker changes several times during the motion, but a detailed analysis is virtually impossible as it depends on too many parameters.
>
> (ii) Finally, we give the solution to the scenario proposed in the *Hint* to this problem. Consider the situation when the student has reached the end of the carriage and stopped, but the ticket collector is still moving backwards with speed v relative to the carriage. In accord with linear momentum conservation, the carriage must be moving forward with speed $u = mv/(M + 2m)$. So, when the student jumps out, he carries away a forwards-directed momentum of $mu = m^2v/(M + 2m)$. When the collector stops, the carriage (with the collector aboard) changes its direction of motion and moves backwards, having a total linear momentum of $-mu$. Thus the final velocity V of the carriage is
>
> $$V = -\frac{m^2}{(M + 2m)(M + m)}v.$$

S71   Taking motion to the right as positive, the initial velocities in the centre of mass frame of the beads (*see* Fig. S71.1) are

$$v_m = \frac{M}{m + M}v_0 \quad \text{and} \quad v_M = -\frac{m}{m + M}v_0.$$

Since their centre of mass is at rest in this frame, the ratio of the two velocities remains constant (at M/m) throughout the motion.

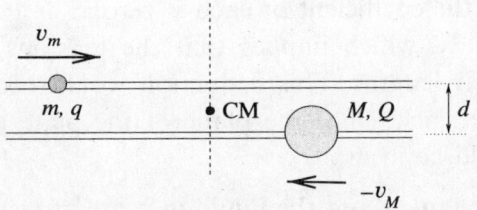

Fig. S71.1

As the beads approach each other, their speeds will decrease if q and Q have the same sign, but increase if their charges are opposite. In the latter case, if d is large enough, their speeds will return to their initial values, since their energies are conserved. In the original frame of reference, after a temporary acceleration, the body of mass m slows until its speed has the original value v_0, while the body of mass M is finally at rest having been displaced (to the left) through a certain distance.

If the beads repel each other, a more detailed discussion is required. If their initial energy is sufficient, the beads pass by each other, and as they part their speeds return to their original values (as viewed from either the centre of mass frame or the 'lab' frame). If, on the other hand, their initial kinetic energy is too low for them to approach within a distance d, they 'turn back'. In the centre of mass system, the body of mass m then moves to the *left* at speed $-v_m$, whilst the body of mass M moves to the *right* with speed $-v_M$ (Fig. S71.2).

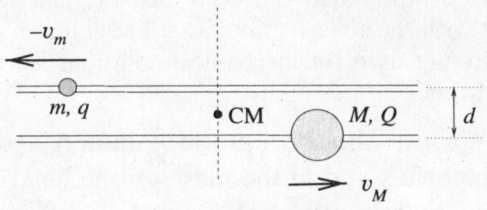

Fig. S71.2

The condition for this to happen is

$$\frac{1}{2}mv_m^2 + \frac{1}{2}Mv_M^2 < \frac{1}{4\pi\varepsilon_0}\frac{qQ}{d},$$

i.e.

$$\frac{1}{2}\frac{mM}{m+M}v_0^2 < \frac{1}{4\pi\varepsilon_0}\frac{qQ}{d}.$$

The quantity $mM/(m+M)$ is called the *reduced mass* of the system. The velocities of the bodies in the laboratory frame can be obtained by adding those in the centre of mass frame to the relative velocity of the two frames; the latter is $|v_M|$. Explicitly,

$$v_m^* = \frac{m-M}{m+M}v_0 \quad \text{and} \quad v_M^* = \frac{2M}{m+M}v_0.$$

In the limiting case, when the initial kinetic energy is just sufficient to allow the two beads to approach within a distance d, the two beads stop with respect to each other (Fig. S71.3), i.e. when viewed in the 'lab' frame, they move on with a common speed $v_m^* = v_M = mv_0/(m + M)$.

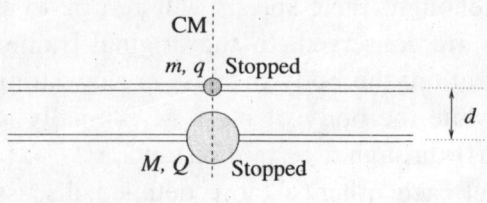

Fig. S71.3

Note. The three cases discussed above model one-dimensional mechanical collisions. The limiting case, when the two bodies move on together, models an inelastic collision. In this case only the mechanical momentum remains constant; the mechanical energy decreases. The velocities in the case in which the bodies approach each other and then move away again are the same as those calculated from the laws of elastic collisions (conservation of energy and momentum). The motion in which the bodies pass each other (in essence, they do not collide and keep their original velocities) is obviously in agreement with the conservation laws. The solution corresponding to this case is usually not used for mechanical collisions since bodies cannot pass through each other.

S72 (i) Let v_0 denote the asymptotic common speed, d the original distance between the beads and m the mass of one bead.

In a given time interval Δt, the cluster collides with $v_0 \Delta t / d$ further beads, which increases its mass by $\Delta m = m v_0 \Delta t / d$ and its momentum by $\Delta p = v_0 \Delta m = m v_0^2 \Delta t / d$. According to Newton's law of motion,

$$F = \frac{\Delta p}{\Delta t} = \frac{m v_0^2}{d},$$

which yields $v_0 = \sqrt{Fd/m}$ for the ultimate speed in the case of inelastic collisions.

(ii) In an elastic collision between two equal mass bodies with one of them initially at rest, their velocities are exchanged. The initially moving body stops, whilst the second one moves away with the same velocity as that initially possessed by the first.

The leftmost bead accelerates uniformly and reaches a speed of

$$v_1 = \sqrt{\frac{2Fd}{m}} = \sqrt{2 v_0}$$

before the first (elastic) collision takes place. It then transfers its speed to the second bead and stops, after which it starts accelerating again as a result of the external force. What happens to the bead it has set in motion? It moves at a constant speed v_1, collides with the third bead and stops. The third and subsequent beads behave similarly, and a 'shock wave' propagates forward at speed v_1.

Meanwhile, the leftmost bead is again accelerated to speed v_1, collides with the second bead, which is now at rest, and the process is repeated, thus starting a new 'shock wave'. The speed of the accelerated bead varies uniformly from zero to v_1, its average value is $v_1/2 = v_0/\sqrt{2} = \sqrt{Fd/(2m)}$.

> *Note.* The case of partially elastic collisions is also interesting. In this case (according to the results of computer simulations) it can be stated that sooner or later, the interacting beads condense into a single cluster that behaves like a perfectly inelastic body with a final speed of $v_0 = \sqrt{Fd/m}$. The time necessary for the cluster to condense depends upon the degree of inelasticity (the coefficient of restitution). The more elastic the elementary collisions, the longer the time necessary for an inelastic cluster to condense.

S73 At any time, the weight of the beer that is in free fall will not be registered by the weighing machine, although the momentum destroyed as the beer is brought to a halt by the jug will be, as will the force experienced by the tap as the direction of the beer flowing through it is changed; there are several such effects to consider. However, if the overall system of table plus jug plus beer is considered these are internal actions and reactions and the only two external forces (ignoring air resistance) are gravity and the upward reaction from the weighing machine. The net result of these two has to be such that the centre of gravity of the system falls, initially accelerating (until the first beer reaches the jug), then sinking with constant velocity, and finally (when the beer runs out) decelerating. Consequently the machine reading, relative to the original, will be: increasingly negative – no change – increasingly positive – no change.

S74 If the cross-section of the incoming water jet is A and its speed is v, then the mass of water of density ρ flowing into the gutter in unit time is of ρAv. This quantity of water has a kinetic energy of $\rho Av^3/2$ and a horizontal momentum of $\rho Av^2 \sin\alpha$. These quantities cannot change if viscosity is neglected, and so

$$\frac{1}{2}\rho Av^3 = \frac{1}{2}\rho A_1 v_1^3 + \frac{1}{2}\rho A_2 v_2^3, \tag{1}$$

$$\rho Av^2 \sin\alpha = \rho A_1 v_1^2 - \rho A_2 v_2^2. \tag{2}$$

The law of conservation of matter has to hold as well, and so we also have

$$\rho Av = \rho A_1 v_1 + \rho A_2 v_2, \tag{3}$$

where A_1 and A_2 are the respective cross-sectional areas of the water flowing out of the gutter to the right and left, while v_1 and v_2 are the corresponding speeds.

Equations (1), (2) and (3) are insufficient to determine the four unknown quantities (the two cross-sectional areas and the two speeds); a further relationship has to be found. According to Bernoulli's law, the quantity $\rho v^2/2 + p + \rho g h$ is constant along a streamline of a non-viscous liquid. Inside the liquid and far from the initial impact point, the pressure is constant and equals the atmospheric pressure. If the difference in the heights of the streams or, more exactly, the change in the energy corresponding to that difference, is neglected (this is correct for a rapidly flowing liquid), the consequence of Bernoulli's equation is that $v = v_1 = v_2$. This means that the liquid leaves the gutter at the same speed at both ends! This is rather surprising, but correct, within the accuracy of the above approximation.

The equations for the conservation of mass and momentum therefore take the forms $A = A_1 + A_2$ and $A \sin \alpha = A_1 - A_2$, which yield

$$\frac{A_1}{A_2} = \frac{1 + \sin \alpha}{1 - \sin \alpha}.$$

This ratio can be examined experimentally and surprisingly good agreement with the calculated values found, which suggests that the approximations made were reasonable.

S75 In a time interval Δt, the level of the liquid with initial acceleration a decreases by $\Delta h = a(\Delta t)^2/2$ and the corresponding mass of liquid which flows out is $\Delta m = (D^2\pi/4)(\Delta h)\rho$. This is equivalent to a decrease of $(\Delta m)gh$ in the potential energy of the liquid as a whole (*see figure*). Meanwhile, the whole of the liquid is accelerated to a speed $\Delta v = a\Delta t$ and its kinetic energy increased by $(D^2\pi/4)h\rho(\Delta v)^2/2$. The speed of the emerging liquid is higher than this, but its effect can be neglected as the quantity of water involved is small compared with the total.

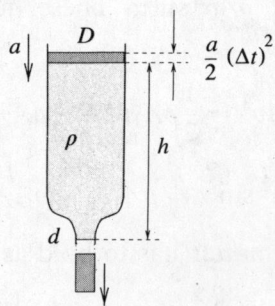

According to the law of conservation of energy, the changes in the potential and kinetic energies are the same:

$$\frac{\pi D^2}{4}\frac{a}{2}(\Delta t)^2\rho gh = \frac{\pi D^2 h\rho}{4}\frac{(a\Delta t)^2}{2},$$

which yields $a = g$.

This means that initially some of the water starts to free fall. According to the mass conservation law, the speed of the emerging water is $(D/d)^2$-times higher than that of the water surface. Consequently, its acceleration must be greater than g by the same factor. For example, if the diameter of the orifice is one-tenth of that of the tube, then the initial acceleration of the emerging liquid is $100g$!

For how long is it true that the liquid is practically in 'free fall'? According to Torricelli's law of efflux, the speed of the efflux is $\sqrt{2gh}$ and the rate of decrease of the liquid level is $(d/D)^2\sqrt{2gh}$. The time interval τ between the start of the efflux and the attaining of (nearly) constant velocities by the surface and the emerging water can be estimated roughly using the relationship

$$g\tau \approx \frac{d^2}{D^2}\sqrt{2gh}.$$

If, for example, $h = 20$ cm and the ratio of the diameters is $1 : 10$ then $\tau \approx 0.002$ s, which is negligible for most purposes.

S76 The sand flows through the aperture almost uniformly, and therefore the total operating time T of the sand-glass is proportional to the volume H^3 of the sand. (As our aim is to obtain only a rough estimate, the difference between the volumes of a cone and a cube is ignored.) The time T may also depend on the gravitational acceleration g, the diameter d of the aperture and the density ρ of the sand, and so $T \approx H^3 \times f(g, d, \rho)$.

As T is a time and only g contains a time dimension, the function f has to be proportional to the reciprocal of the square root of g. Similar reasoning shows that T cannot depend on ρ, but is proportional to $d^{-5/2}$; in summary, $T \approx H^3/\sqrt{d^5 g}$. The coefficient of proportionality is a dimensionless number, and since it does not depend on anything, can be assumed to be of order 1 (though such assumptions are notoriously dangerous in some branches of physics!).

Consider some realistic data. If, for example, H is a few centimetres and d is around a millimetre, T is a few minutes, which is indeed the sort of time for which an egg should be boiled.

Note. In principle, the average diameter of the grains of sand could be included in the formula. However, this is in general much smaller than the other dimensions involved, therefore (in the same way as the size of the atoms constituting the sand is ignored) it does not play a role in the dimensional reasoning. The trickiest point in all dimensional analysis is that of choosing the relevant parameters for the phenomenon in hand.

S77 Let the displacement of the bob be x and let us calculate the net force (F) exerted on it (*see* Fig. S77.1).

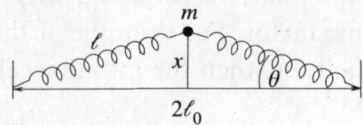

Fig. S77.1

The length of the extended spring is

$$\ell = \sqrt{\ell_0^2 + x^2} \approx \ell_0 + \frac{x^2}{2\ell_0},$$

and so the tension in it is

$$F_{\text{spring}} = k\,\frac{x^2}{2\ell_0}.$$

The *net* force acting on the bob (*see* Fig. S77.2) is

$$F = -2F_{\text{spring}} \sin\theta \approx -2F_{\text{spring}}\frac{x}{\ell_0} = -\frac{k}{\ell_0^2}x^3,$$

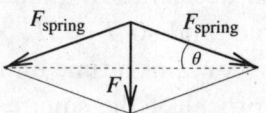

Fig. S77.2

and the resulting equation of motion is

$$m\frac{d^2x}{dt^2} = -\frac{k}{\ell_0^2}x^3.$$

This is a differential equation which cannot be solved by elementary methods.

However, to solve the problem as posed it is not necessary to solve the equation explicitly, only to apply dimensional analysis to it. Writing the equation in the form

$$\frac{d^2x}{dt^2} = -\frac{k}{m\ell_0^2}x^3 = -Cx^3,$$

we can suppose that the period of the motion T, depends only on C and A, the amplitude of the vibration. The dependence can be written as

$$T \propto C^\alpha \times A^\beta,$$

implying the dimensional relationship

$$s = [T] = [C]^\alpha \times [A]^\beta = \left(\frac{kg}{s^2} \times \frac{1}{kg} \times \frac{1}{m^2}\right)^\alpha \times m^\beta = s^{-2\alpha} \times m^{-2\alpha+\beta}.$$

This equation is satisfied if $-2\alpha = 1$ and $-2\alpha + \beta = 0$. Therefore $\alpha = -\frac{1}{2}$ and $\beta = -1$, implying that $T \propto 1/A$. Accordingly when the amplitude is doubled (2 cm), the period is halved (1 s).

Since for dimensional analysis the choice of variables to include appears more or less arbitrary, thus casting some doubt on the validity of the conclusions reached, we now give another method for the solution of the current problem.

The velocity of the bob as a function of its position can be calculated from the law of conservation of energy. The stored energy in the two springs when the bob is at rest at its maximum displacement must be equal to the sum of the kinetic energy of the bob and the stored spring energy for a general displacement. As the elongation of each spring is

$$\Delta\ell = \frac{1}{2}\frac{x^2}{\ell_0},$$

the equality may be written as

$$\frac{kA^4}{4\ell_0^2} = 0 + 2 \times \frac{1}{2}k\left(\frac{A^2}{2\ell_0}\right)^2 = 2E_{\text{spring}}^{\text{max}} = \frac{1}{2}mv^2 + 2 \times \frac{1}{2}k\left(\frac{x^2}{2\ell_0}\right)^2,$$

where A is the maximum displacement (amplitude) of the bob. From this equation the velocity of the bob can be expressed as

$$\frac{dx}{dt} = v = \frac{1}{\ell_0}\sqrt{\frac{k}{2m}}\sqrt{A^4 - x^4}.$$

After separating the variables we get

$$\ell_0\sqrt{\frac{2m}{k}}\int_0^A \frac{dx}{\sqrt{A^4 - x^4}} = \int_0^{T/4} dt = \frac{T}{4}.$$

Using $y = x/A$ as a new variable, our final result becomes

$$T = \ell_0 \sqrt{\frac{32m}{k}} \frac{1}{A} \int_0^1 \frac{\mathrm{d}y}{\sqrt{1 - y^4}}.$$

This expression shows directly that the period is inversely proportional to the amplitude, again leading to the conclusion that doubling the amplitude of the motion to 2 cm, reduces its period to 1 s.

> *Note.* The definite integral in the final expression for the period can be evaluated using no more than a programmable calculator:
>
> $$\int_0^1 \frac{\mathrm{d}y}{\sqrt{1 - y^4}} \approx 1.31,$$
>
> thus giving the complete solution to the problem.

S78 Because of the weakness of the spring, the body falls virtually freely at first. The length of the spring is soon several times larger than its unstretched length (which can be neglected during subsequent motion). With this approximation, the body executes simple harmonic motion, both vertically and horizontally. As it is released with no initial speed, it arrives vertically under the suspension point after a quarter of the period of the horizontal motion. Meanwhile, the vertical motion has also completed a quarter-cycle, and the body has sunk to its equilibrium position at a depth of mg/k (this is much larger than L).

The motion can be described quantitatively. In the coordinate system shown in the figure, the equations of motion of the body at point (x, y) are:

$$ma_x = -k \left(\sqrt{x^2 + y^2} - L \right) \frac{x}{\sqrt{x^2 + y^2}},$$

$$ma_y = -k \left(\sqrt{x^2 + y^2} - L \right) \frac{y}{\sqrt{x^2 + y^2}} + mg.$$

During the first part of the motion, whilst the extension of the spring is not much larger than L, the force exerted on the spring can be neglected. On the other hand, when

$$\sqrt{x^2 + y^2} \gg L$$

the original length of the spring can be neglected and the equations of motion take the following simple forms:

$$ma_x = -kx \quad \text{and} \quad ma_y = -ky + mg.$$

These equations describe harmonic oscillations of identical periods, about the origin in the x-direction and about the equilibrium position $y_0 = mg/k$ in the y-direction. Incorporating the initial conditions gives the solution:

$$x(t) = L\cos\left(\sqrt{\frac{k}{m}}t\right), \qquad y(t) = \frac{mg}{k}\left[1 - \cos\left(\sqrt{\frac{k}{m}}t\right)\right].$$

The body is under the point of suspension when $x(t) = 0$, and $y = y_0 = mg/k$, in agreement with our previous conclusion.

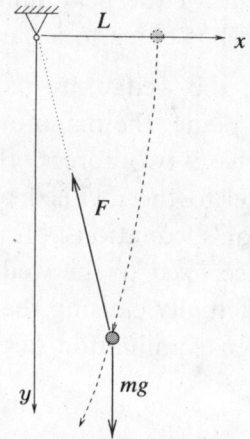

Note. For the early part of the motion $t \ll \sqrt{m/k}$ (when the assumed equations of motion are not strictly valid), the above expressions for x and y can be approximated by $x(t) \approx L$ and $y(t) \approx gt^2/2$, which are in agreement with the formulae describing free fall, as is appropriate to that part of the motion.

S79 If the carriage brakes with deceleration a, then in the carriage reference frame, a 'virtual inertial force' of magnitude ma, in the direction of the carriage's motion, will appear to act on the body.

If this inertial force acted permanently, the pendulum could certainly not reach the vertical, since, if it did, the net work done by the inertial force would be zero (the net displacement of its point of application would be perpendicular to its line of action) and the gravitational force would be negative, implying that the kinetic energy of the pendulum should be negative. This is impossible.

Consider now the fact that the carriage only brakes for a certain length of time (until it stops). If it stops when the thread of the pendulum is horizontal, the work done by the inertial force is $W = maR$, where R is the length of the thread. If the pendulum subsequently reaches a vertical position with speed

v then, from the conservation of energy,

$$maR - 2mgR = \frac{mv^2}{2}.$$

For the thread to remain taut, even at the topmost point, requires $mv^2/R >$ mg, which, together with the above relation for the velocity, implies that the deceleration of the railway carriage $a > 2.5g$. The conclusion is, therefore, that the taut thread can reach the vertical provided the deceleration is great enough and v_0 is large enough for the pendulum to have time to reach the horizontal before the carriage has come to a halt.

S80 The forces acting on the wedge are its weight mg and a force K, perpendicular to the inclined plane; the magnitude of the latter may change with time. As a result of these two forces, the only component of the wedge's acceleration **a** parallel to the inclined plane is $g \sin \alpha$ (as measured in an inertial frame). Newton's equations of motion remain valid in an accelerating frame of reference fixed to the wedge only if an 'inertial force' $-m'\mathbf{a}$ is added to the forces actually causing the motion of a body. Here m' is the mass of the body under examination (e.g. that of a small volume of water).

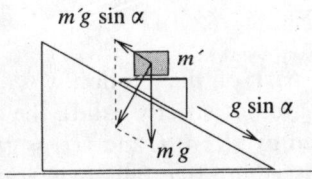

The resultant of the gravitational and inertial forces acting on the mass m' must be perpendicular to the inclined plane as the components parallel to it cancel each other. The bodies on the wedge (the glass and the water in it) 'feel' as if they were in a gravitational field perpendicular to the inclined plane, with the consequence that the surface of the water lies parallel to the plane.

This statement does not depend on the motion of the plane; it can be fixed or move freely or even – as the result of a small force – be shaken to and fro. As long as the friction between the inclined plane and the wedge is negligible and the wedge does not rise off the plane, the shape of the water surface cannot be other than a plane parallel to the inclined surface.

The case $m \gg M$ deserves an additional comment. In this case, the wedge 'pushes away' the inclined plane, and falls nearly freely. The weight of the bodies on the wedge (including the water) are nearly completely 'lost'; but even so, the small force keeping the water inside the glass is still sufficient to set the water surface parallel to the inclined plane.

> *Note.* The water surface would only become parallel to the inclined plane after a long time, and on a correspondingly long plane. This is why this interesting phenomenon cannot be observed experimentally in normal circumstances.

S81 Assume that the string is of uniform cross-section and mass distribution, and is free at both ends. It orbits the Earth in such a way that its position relative to the Earth is always the same. Obviously, if the string is in a vertical position, the phenomenon could only occur at the Equator.

In the Earth's reference frame a body of mass m orbiting above the Equator at a distance r and with angular velocity ω experiences a gravitational force of $-GMm/r^2$ and a centrifugal force of $mr\omega^2$. Here M is the mass of the Earth and G is the gravitational constant. The condition for the equilibrium of the string is that the net force due to gravitation, which varies with r, is equal to that due to the centrifugal effect, which also changes from point to point. This condition can easily be derived using integral calculus, but it can also be found without using such sophisticated mathematics.

Imagine that the string is pulled down a little by some external force. Since (in the rotating frame of reference) the string was initially in equilibrium, it can be displaced from its equilibrium position by an arbitrarily small force and, to first order, the net work done in the course of the change must be zero. The displacement of the whole string – from the point of view of the work done – is equivalent to the slow migration of a small piece of the string, of mass Δm, from its top to its bottom. The work done is the sum of two terms, the change in the gravitational potential energy and the work done by the average centrifugal force (since the centrifugal force changes linearly). If the bottom end of the string of length L just touches the Earth's surface, the work in question is

$$W = GM\Delta m \left(\frac{1}{R} - \frac{1}{L+R} \right) - \Delta m \frac{R + (R+L)}{2} \omega^2 L = 0,$$

where R is the radius of the Earth. This is a quadratic equation in L, which gives

$$L = \frac{R}{2} \left(-3 + \sqrt{1 + \frac{8GM}{R^3\omega^2}} \right) \approx 140\,000 \text{ km}$$

using known data. This length is several times $r_s = (GM/\omega^2)^{1/3} \approx 42\,000$ km, the distance of telecommunications satellites from the centre of the Earth!

What is the maximum stress in the string? It is easy to establish that the greatest stress σ_{max} occurs at the position $r = r_s$, and satisfies $\sigma_{max}/\rho = 4.8 \times 10^7$ N m kg^{-1}, where ρ is the mass density of the string. This figure for the ratio of tensile strength to density is much greater than that for known materials (for steel it is 2.6×10^5, and for carbon 1.7×10^6). Therefore, whilst, in principle, a 'hook to the sky' is consistent with Newton's laws, at the present time it is impossible (at least with a string of constant cross-section) to find suitable materials from which to build it.

S82 (i) At the highest point of the bridge the equation of motion of the car is

$$mg - N = m\frac{v^2}{\rho},$$

where N is the normal force acting on the car (and the negative of the required answer), $v = 20$ m s^{-1} and ρ is the radius of curvature of the bridge there. The most difficult part of the problem is to find this radius of curvature.

If we could find a motion with this trajectory for which the normal acceleration is well known, the radius of curvature could be easily calculated. For a parabolic trajectory the flight of a projectile offers the required analogue. Let the projectile have an initial velocity of v_0 making an angle α with the horizontal.

The range ($d = 100$ m) and height ($h = 5$ m) of the projectile can be expressed using the initial data,

$$d = \frac{2v_0^2 \sin \alpha \cos \alpha}{g} \quad \text{and} \quad h = \frac{v_0^2 \sin^2 \alpha}{2g}.$$

The quotient h/d gives $\tan \alpha = 4h/d$ (so $\alpha \approx 11.3°$), and the horizontal component of the initial velocity is

$$v_x = v_0 \cos \alpha = d\sqrt{\frac{g}{8h}} = 50 \text{ m s}^{-1}.$$

Now the radius of curvature at the highest point can be calculated as $\rho = v_x^2/g = 250$ m.

So the normal force at the highest point is

$$N = m\left(g - \frac{v^2}{\rho}\right) = 8.40 \text{ kN}.$$

(ii) The force exerted on any other part of the bridge can be calculated in the same way, i.e. using the radius of curvature. At a point three-quarters of the way across the bridge, the radius of curvature is approximately 254 m and the normal force about 8.37 kN. Away from the centre of the bridge there is also a tangential (frictional) force; here its value is 995 N, and so the net force acting on the bridge is approximately 8.43 kN.

S83 We can prove that the radii of curvature of the ellipse at the endpoints of its axes are b^2/a and a^2/b, where $2a$ and $2b$ are the lengths of the major and minor axes, respectively. This geometrical result can be deduced using calculus or by considering one of a number of physical situations; what follows is one possibility.

Consider a planet orbiting the Sun in an ellipse. Newton's second law of motion applied at the endpoint of the major axis, a distance r from the Sun, gives

$$G\frac{M}{r^2} = \frac{v^2}{R},$$

where R is the radius of curvature at the endpoint and M is the mass of the Sun. According to Kepler's third law the period of the orbit is $2\pi\sqrt{a^3/GM}$ and the radius vector sweeps out area at a constant rate. The area of the ellipse is πab, and so equating two expressions for that constant rate when the planet is at the endpoint of the major axis, we obtain

$$\frac{vr}{2} = \frac{ab}{2}\sqrt{\frac{GM}{a^3}}.$$

Comparing the above two equations we conclude that $R = b^2/a$. For this argument we utilised the fact that the foci of the ellipse are on the major axes; we cannot therefore apply the same proof at the endpoints of the minor axis. However, in respect of their corresponding radii of curvature, the two axes are symmetrical.

The uniformly moving point mass of the problem obeys the equation of motion $F = mv^2/R$, where R is the appropriate radius of curvature. Using the data given we obtain: $b^2/a = 1.25$ m; $a^2/b = 10$ m and, hence, $2a = 10$ m; $2b = 5$ m.

> *Note.* The radii of curvature of the ellipse could also be calculated using well-known formulae from SHM. Consider the point mass moving in the x–y plane around an ellipse with semi-axes a and b according to the equations
>
> $$x = a\cos\omega t \quad \text{and} \quad y = b\sin\omega t.$$

At $t = 0$ the mass is moving at the end of the major axis with velocity $v = b\omega$ and acceleration $A = a\omega^2$. On the other hand, the acceleration is $A = v^2/R$; so the radius of curvature is $R = b^2/a$. Similarly, we find the radius of the curvature at the end of the minor axis to be a^2/b.

S84   Denote the width of the canal by d and draw a straight line perpendicular to its banks a distance d downstream from the boat's starting point A (*see figure*).

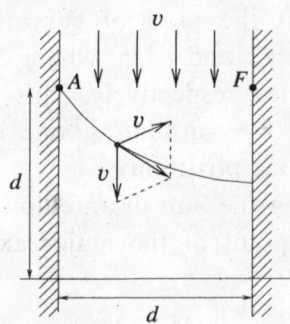

The boat is initially at distance d both from the mark F on the opposite bank and from this straight line. As both the speed of the water and that of the boat with respect to the water are v, the water takes the boat downstream by the same distance as is covered by the boat in the direction of F.

This means that the boat is always equally far from point F and the straight line. The path of the boat is therefore a parabola, with F as its focus and the straight line as its directrix. After a very long time, the boat approaches the opposite bank at a point $d/2$ from F. Because the speed of the current equals that of the boat, the boatman cannot land closer than this.

S85   If, after the slightest of pushes, the child would slide (straight) downhill at a steady speed, the component F of its weight parallel to the inclined plane must have the same magnitude as the frictional retarding force S, i.e. $F = S$.

The force of kinetic friction – the direction of which is always opposite to that of the instantaneous velocity – causes the speed to decrease, while the force F increases the component of the velocity parallel to the inclined plane. These two effects are of course present together and result, in general, in a rather complicated motion (on a curved path and with a changing acceleration). Despite this, the final speed can be determined without the need for a detailed description of the motion.

The figure shows a coordinate system for the general situation in which the child's trajectory is not straight downhill. Denote the magnitude of the instantaneous velocity of the sliding child by v, and its component in the y-direction by v_y. We first calculate the change in these two quantities in a short time interval Δt. According to Newton's second law:

$$m\Delta v = (-S + F \cos \alpha)\Delta t, \qquad m\Delta v_y = (F - S \cos \alpha)\Delta t.$$

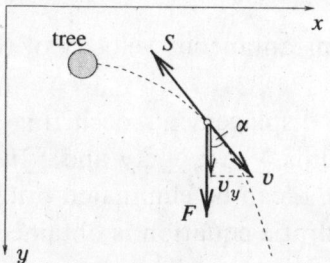

Adding these two equations and using $F = S$ gives

$$\Delta v + \Delta v_y = \Delta(v + v_y) = 0,$$

i.e.

$$v + v_y = \text{constant}.$$

From the initial conditions, the value of this constant is $v_0 = 1 \text{ m s}^{-1}$. The final speed v_{max} of the sliding child is directed down the slope, and its magnitude is determined by the above 'conservation law' with $v = v_y = v_{max}$, i.e.

$$v_{max} = \frac{v_0}{2} = 0.5 \text{ m s}^{-1}.$$

S86 Let kv denote the speed of the coastguard's cutter, i.e. k is the required ratio of the speeds of the two vessels.

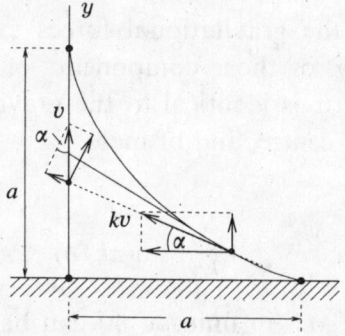

At a general time t, as shown in the figure, the distance d between the ships (initially a) decreases by

$$\Delta d = kv\Delta t - v \sin \alpha \, \Delta t \tag{1}$$

in time Δt. Meanwhile, the distance of the cutter from the shore increases by

$$\Delta y = kv \sin \alpha \, \Delta t, \tag{2}$$

where α is the angle the instantaneous velocity of the cutter makes with the shore.

We now sum the small displacements occurring in equations (1) and (2), knowing that the three sums $\sum \Delta d$, $\sum \Delta y$ and $\sum v\Delta t$ must all equal a. The potentially awkward angle α can be eliminated prior to the summations and a surprisingly simple quadratic equation is obtained for k,

$$k^2 - k - 1 = 0, \quad \text{with} \quad k = \frac{1 + \sqrt{5}}{2} \approx 1.618$$

as its positive root. This figure is the famous 'golden mean' associated with the Fibonacci series. In the current situation, it is the ratio of the speeds of the coastguard's cutter and the smugglers' ship if they are to meet as described in the problem.

S87 From the symmetry of the layout and initial conditions, we deduce that all the bodies fall towards the centre of the n-gon with the same non-uniform acceleration. The formation keeps its original shape, but the distance r from the centre decreases at a non-uniformly accelerating rate. The resultant force acting on one (say the nth) body when it is at distance r from the centre is

$$F(x) = G\frac{m^2}{r^2} \sum_{k=1}^{n-1} \frac{1}{4\sin(\pi k/n)}.$$

This force, made up of the gravitational forces exerted by all the other bodies, or, more precisely, of those components of these forces which are directed towards the centre, is identical to the gravitational attraction of a fixed body situated at the centre, and of mass

$$M_n = \frac{m}{4} \sum_{k=1}^{n-1} \frac{1}{\sin(\pi k/n)}.$$

The values of the masses M_n (in units of m) can be calculated numerically for all values of n as

$$M_2 = 0.25, \quad M_3 = 0.58, \quad M_4 = 0.96, \quad \dots, \quad M_{10} = 3.86, \quad \dots$$

The time T of the collapse from an initial distance R onto a central mass M can be considered as half of the period T_e for a severely flattened (degenerate) elliptical orbit of major semi-axis $R/2$. The period T_c of a circular orbit of radius R can be calculated directly from the dynamical equation for circular motion,

$$G\frac{Mm}{R^2} = mR\left(\frac{2\pi}{T_c}\right)^2, \qquad \text{giving } T_c = 2\pi\sqrt{\frac{R^3}{GM}}.$$

But, according to Kepler's third law,

$$\left(\frac{T_e}{T_c}\right)^2 = \left(\frac{R/2}{R}\right)^3.$$

Thus, finally, we obtain $T = \pi\sqrt{R^3/8GM_n}$ for the time.

> *Note.* The limiting case $n \gg 1$ is interesting. As the number n of bodies increases, M_n increases even if the total mass of the system is fixed at M_0, i.e. $m = M_0/n$. The more finely a given amount of matter is spread around a circle, the shorter the time it takes for it to collapse under its own gravitational attraction. However, there is no point in examining a continuous matter distribution spread along an arbitrarily thin line; the extent of the matter in the transverse, i.e. radial, direction cannot be neglected.

S88 According to Kepler's first law the orbit of the rocket is an ellipse with one of its foci at the centre of the planet. The launch and return velocities are parallel to each other (though in opposite directions) if the launch and return points are at the ends of the minor axis of the ellipse. But, for an ellipse, the distance from a focus to either end of the minor axis is equal to the length a of its major semi-axis; consequently $a = R$ (see Fig. S88.1).

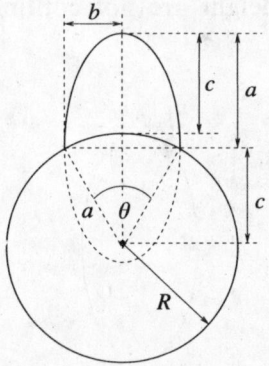

Fig. S88.1

From Kepler's third law, satellites in orbits having different eccentricities, but the same lengths of major axis, have equal periods, and so in our case the period for a full orbit would be the given T_0. The rocket, however, covers only one-half of the ellipse. The time required for this is not half of the full period, but proportional to the fractional area swept by the radius vector joining the rocket to the focus (Kepler's second law). The area of the whole ellipse is

$$A_0 = \pi ab = \pi a^2 \sin \frac{\theta}{2}.$$

The swept area for the half orbit is

$$A_1 = \frac{\pi ab}{2} + \frac{1}{2} \times 2bc = \frac{1}{2}a^2\pi \sin \frac{\theta}{2} + a^2 \sin \frac{\theta}{2} \cos \frac{\theta}{2}.$$

So the flight time is

$$T_1 = \frac{A_1}{A_0} T_0 = T_0 \left(\frac{1}{2} + \frac{1}{\pi} \cos \frac{\theta}{2} \right).$$

The maximum distance above the surface of the planet is

$$2a - a - (a - c) = c = R \cos \frac{\theta}{2} \leq R.$$

If the angle between the launch and arrival points is allowed to approach zero ($\theta \to 0$), the calculated flight time approaches a maximum value of

$$T_0 \left(\frac{1}{2} + \frac{1}{\pi} \right),$$

and the maximum height achieved approaches the radius of the planet ($c \to R$). But, in fact, if the take-off and landing sites are the same ($\theta = 0$), the rocket can reach any arbitrary height, large or small. This implies that the period and maximum height are not continuous functions of θ at the point $\theta = 0$.

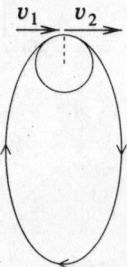

Fig. S88.2

If the launch speed is sufficiently great (equal to or larger than the first cosmic speed, $v = \sqrt{Rg}$) and the initial velocity is tangential to the surface of the planet, then the orbit shown in Fig. S88.2 is possible. Again the return velocity is parallel to the launch one, but this time in the same direction. The maximum height achieved can be anything, but the period must be at least T_0. These are the orbits corresponding to the special case $\theta = 0$.

S89 (i) Writing κ for $\mu_0/4\pi$, the couple on C due to D is $\kappa\mu^2/L^3$ anti-clockwise; that of C on D is $2\kappa\mu^2/L^3$, also anti-clockwise.

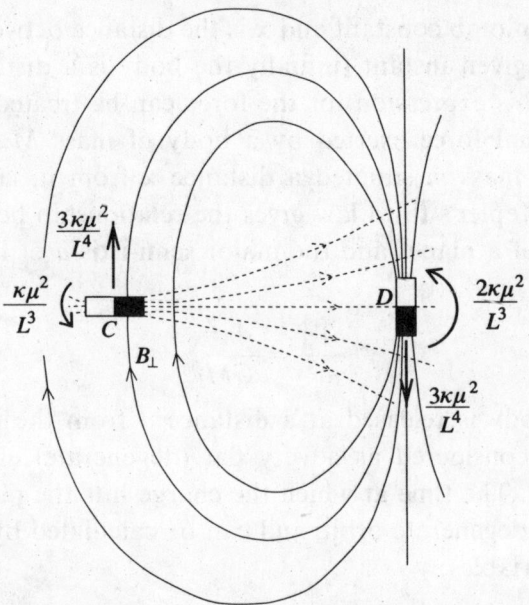

(ii) Couples are not the only potential result of the magnetic fields; forces will result if a dipole is positioned in a non-uniform magnetic field. The force on C due to D is non-zero as the strength of $B_\perp$ is slightly less at the position of one of the poles of C than at the other. The magnitude of the net force is $\mu \times \partial B_\perp/\partial r$ where the derivative is evaluated at $r = L$, and is thus $3\kappa\mu^2/L^4$; its direction is the same as that of $B_\perp$. This force and its reaction on D produce a couple on the rod which has magnitude $3\kappa\mu^2/L^3$ and acts in the clockwise sense. It thus exactly cancels the other two couples acting on the rod and when the system is suspended, nothing at all happens – something that must be clear on the grounds of symmetry and the impossibility of free perpetual motion!

> *Note.* This is an unusual example of a non-central force and its reaction, which act along parallel, but not identical, lines.

S90 The charge distribution induced on the plane by the charge q, produces (in the region above the plane) an electrostatic field identical to that of a charge $-q$ situated below the plane at the point which is the mirror image of the body's position, if the plane were considered as a mirror. This is the principle of image charges. Thus, the force of attraction acting on the body (moving non-relativistically) can be calculated using Coulomb's law as

$$F(x) = k\frac{q^2}{4x^2},$$

where k is the Coulomb constant and x is the distance between the plane and the body at any given instant (initially the body is a distance $2d$ from the image charge). This expression for the force can be treated as the analogue of the gravitational force exerted by a body of mass $M = kq^2/(4Gm)$ on another body of mass m situated a distance x from it, i.e. $F = GMm/x^2$. In this analogy, Kepler's third law gives the relationship between the period of revolution T of a planet and the major semi-axis a of its elliptical orbit, namely,

$$\frac{T^2}{a^3} = \frac{4\pi^2}{GM}.$$

If the charged body is released at a distance d from the metal plane then its orbit can be considered as a very flat (degenerate) ellipse with major semi-axis $a = d/2$. The time at which the charge hits the plane T_h, is half of the period of the degenerate orbit, and can be calculated by substituting the corresponding variables:

$$T_h = \frac{T}{2} = \frac{\pi}{q}\sqrt{\frac{md^3}{2k}}.$$

S91   Brine is a good conductor, because positive and negative ions can move easily within it. When the charged plastic ball is placed close to the surface of the water, opposing charges are induced in the surface, whilst like charges are repelled from it. The resulting electric field lines above the water surface will be perpendicular to it, whilst beneath it the net electric field vanishes.

The charged ball attracts the water below it, and the surface wells up in a hump. The electrical forces exerted on the hump are balanced mainly by gravity and the effect of surface tension can be ignored. We don't know the shape of the hump exactly, but can be sure that the rise in water level will be small and there will be only a slight deviation from a plane surface; this is why we can use the so-called method of image charges. It will be sufficient to consider the maximum effect and find the rise at the point P shown in Fig. S91.1.

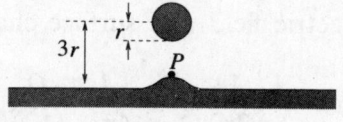

Fig. S91.1

At P the electric field due to the charge Q is

$$E_1 = \frac{1}{4\pi\varepsilon_0} \frac{Q}{(3r)^2}.$$

The effect of the unknown surface charge distribution can be replaced by that due to an image charge of $-Q$ situated at a depth below the surface of $3r$ (*see* Fig. S91.2). The electric field at P due to the image charge has the same magnitude and direction as E_1, and so the net electric field is

$$E = 2E_1 = \frac{1}{2\pi\varepsilon_0} \frac{Q}{(3r)^2}.$$

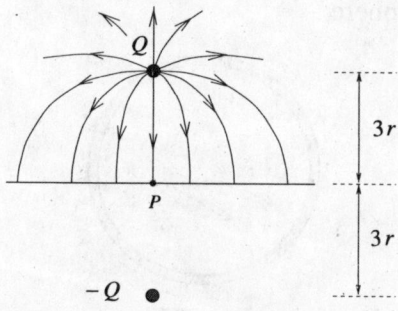

Fig. S91.2

According to Gauss's law the surface charge density at P is

$$\sigma = \varepsilon_0 E = \frac{1}{2\pi} \frac{Q}{(3r)^2}.$$

At the water surface the force exerted on a unit area is the product of the surface charge density σ and the electric field E_1 due to the ball:

$$\frac{F}{A} = \sigma\, E_1.$$

This is the upward force at P per unit area and is balanced by the hydrostatic pressure associated with the maximum rise h in water level:

$$\frac{F}{A} = \rho g h.$$

Substituting for the electric field and surface charge gives an expression for h as

$$h = \frac{1}{\rho g} \frac{1}{2\pi} \frac{Q}{(3r)^2} \frac{1}{4\pi\varepsilon_0} \frac{Q}{(3r)^2}.$$

Inserting the numerical data into this equation, yields $h \approx 0.29$ mm, which is very small compared to the diameter of the ball and justifies our treating the water surface as being close to flat.

S92 The method of spherical image charges can be applied. Let two point charges of opposite signs be $+Q_1$ and $-Q_2$. In the field produced by them the locus of points of zero potential is given by

$$k\frac{Q_1}{r_1} - k\frac{Q_2}{r_2} = 0,$$

where r_1 and r_2 are the distances from the two charges of a general point on the locus. A straightforward rearrangement gives $Q_1/Q_2 = r_1/r_2$, i.e. the distance ratio r_1/r_2 is constant. According to Apollonios's theorem, points with this property lie on a sphere (the Apollonios sphere). Therefore the zero potential surface is a sphere.

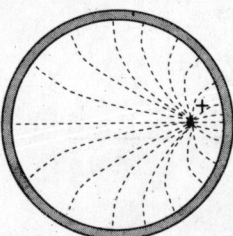

Fig. S92.1

If the spherical metal shell mentioned in the problem is earthed, then it is at zero potential. The point charge $+Q$ inside it induces an inhomogeneous charge distribution on the inner surface of the shell as shown in Fig. S92.1. The electric field inside the shell, due to the actual charge and the induced charge distribution, is the same as if it were caused by the actual charge and a negative point charge outside the spherical shell. The latter is what is called a spherical image charge.

The charge distribution inside the spherical metal shell is independent of the potential of the shell. If the shell were not earthed then charge $+Q$ would appear uniformly distributed over its outer surface, regardless of the inner charge distribution. This is because the electric field strength is zero everywhere inside the material of the shell and therefore the charge on the outer surface is not aware of the presence of that on the inner one. The external electric field makes it appear as if the enclosed charge were at the centre of the sphere.

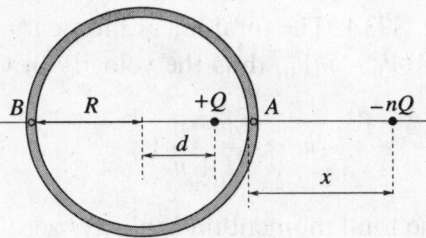

Fig. S92.2

The force acting on the charge Q inside the shell is equal to the Coulomb force acting between the charge and the corresponding image charge. Using the notation in Fig. S92.2, the charge $+Q$ is a distance d from the centre of the sphere of radius R, while the image charge $-nQ$ is a distance x from the shell. The ratio of the charges is therefore n, an expression for which can easily be found using the two points A and B in which the straight line connecting the charges intersects the spherical shell:

$$n = \frac{x}{R - d} = \frac{x + 2R}{R + d}.$$

This yields for n and x that $n = R/d$ and $x = R(R - d)/d$. The force acting on the charge inside the spherical metal shell is therefore

$$F = -k \frac{nQ^2}{(x + R - d)^2} = -kQ^2 \frac{Rd}{(R^2 - d^2)^2}.$$

It is clear that this force is zero when $d = 0$ and tends to infinity as $d \to R$. The negative sign shows that it is directed towards the position of the (imaginary) image charge.

> *Note.* The electric charge distribution on the inner surface of the spherical metal shell can be calculated using Gauss's law. The magnitude of the surface charge density is proportional to the electric field strength obtained by superimposing the fields of the real and image charges.

S93 Denote the mass of the boron atoms (actually boron ions) by M and that of the unknown colliding particles by m.

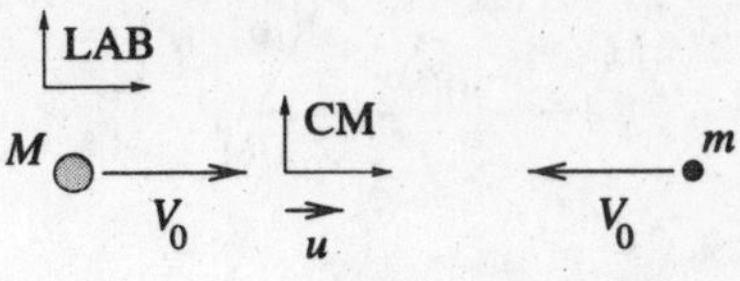

Fig. S93.1

Before the collision the particles have opposing velocities of the same magnitude V_0 as measured in the laboratory (LAB) frame of reference. We can easily transform to the centre of mass (CM) frame of the colliding particles shown in Fig. S93.1. The total linear momentum of the two particles in the LAB frame is $MV_0 - mV_0$, thus the velocity of CM in this frame is

$$u = \frac{M - m}{M + m}V_0.$$

In the CM frame the total momentum is always zero and the two particles must always move in opposite directions with linear momenta of equal magnitudes. However, in accord with conservation of energy, the magnitude of the momentum and therefore the velocity of each particle must be the same before and after the collision – only their directions can change. The speed of the boron atoms in the CM frame before the scattering is

$$V = V_0 - u = V_0 - \frac{M - m}{M + m}V_0 = \frac{2m}{M + m}V_0,$$

and so it must also have this value after the collision (Fig. S93.2).

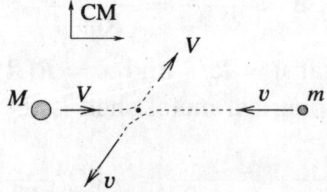

Fig. S93.2

We can return to the LAB frame by adding **u**, the relative velocity of the frames, to the CM velocity vectors. In the LAB frame the velocity of the boron atom after scattering $\mathbf{u} + \mathbf{V}$, is a vector pointing to some point on the circle shown in Fig. S93.3. The maximum angle between the final and initial velocities of the boron atom occurs if $\mathbf{u} + \mathbf{V}$ is tangential to this circle of radius **V**, i.e. **V** is perpendicular to $\mathbf{u} + \mathbf{V}$.

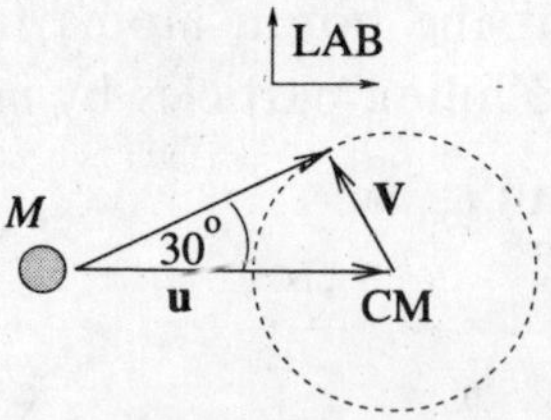

Fig. S93.3

In this case

$$|\mathbf{V}| = \sin 30° |\mathbf{u}|, \qquad \text{i.e.} \qquad \frac{2m}{M+m}V_0 = \frac{1}{2}\frac{M-m}{M+m}V_0.$$

This gives $m = \frac{1}{5}M$. Thus the unknown particle has a mass number of $A = 2$, and is actually not unidentified any longer; the particle is the *deuteron*, the nucleus of deuterium.

S94 Friction between the two balls is negligible, and so, during the collision, they can only exert forces normal to their surfaces. Thus, the first ball stops after the collision, while the second acquires the first's initial speed v_0. The rotation of the balls, however, does not change, and so, immediately after the collision, the first ball rotates on one spot and the second slides without rotation at speed v_0.

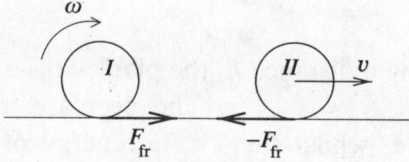

Fig. S94.1

The friction between the balls and the table is of course important and affects the motion of the balls. The first is accelerated forward by the force of kinetic friction $F_{fr} = \mu m g$, whilst the second is slowed by the same force, as shown in Fig. S94.1. The rotation of the first ball is reduced by friction, that of the second one is increased. The part played by the frictional force lasts until both balls reach the state of rolling without slipping. After that their motion is unchanged.

It will be shown that the final motion of the balls depends neither on the frictional coefficient, nor on the possible variation of it with position. After the collision, the initially moving ball rotates with an angular velocity $\omega = v_0/r$. Its angular momentum about its axis is therefore $I\omega = \frac{2}{5}mr^2(v_0/r) = \frac{2}{5}mv_0 r$. The angular momentum about the point of contact with the table P, must be the same, since the centre of mass of the ball is at rest, i.e. the angular momentum attributable to translation is zero. The angular momentum of the ball about P cannot be changed by friction any more, as the line of action of this force runs through P. (The sum of the gravitational force and the reaction of the table is zero, and so they can produce no net torque either.)

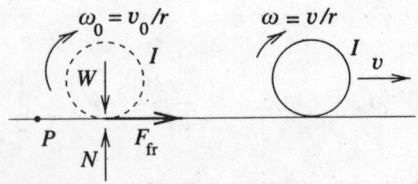

Fig. S94.2

The angular momentum of a ball, rolling without slipping at speed v, is the sum of its own angular momentum $\frac{2}{5}mvr$ and the angular momentum mvr due to the motion of its centre of mass. Figure S94.2 shows (on the left) the initially moving ball and the forces acting on it shortly after the collision. On the right of the figure, the ball is shown ultimately rolling without slipping. According to the law of conservation of momentum $\frac{2}{5}mv_0r = \frac{2}{5}mvr + mvr$, which yields $v = \frac{2}{7}v_0$. Similar reasoning shows that the final speed of the other ball has to be $\frac{5}{7}v_0$, regardless of the magnitude of the coefficient of friction.

S95   When travelling a distance L, the plank causes L/d rollers to acquire an angular velocity $\omega_{max} = v_{max}/r$. The decrease in potential energy of the plank is $MgL \sin \alpha$, whilst the kinetic energy of each roller becomes $\frac{1}{2}I\omega_{max}^2 = \frac{1}{4}mv_{max}^2$. Notice that the final tangential surface speed of each roller is equal to the terminal speed of the plank, and the moment of inertia of each roller is $I = \frac{1}{2}mr^2$.

It is false reasoning to suppose that the lost gravitational potential energy of the plank is simply converted into kinetic energy of the rollers. This would lead to concluding from the equation

$$MgL \sin \alpha = \frac{L}{d}\frac{1}{4}mv_{max}^2 \tag{1}$$

that the terminal speed of the plank is

$$v_{max} = \sqrt{\frac{4dMg \sin \alpha}{m}}.$$

However, this result is *wrong*, because it does not take into account the fact that the speeding-up of the rollers involves kinetic friction and, consequently, there is heat dissipation in the process.

Denote by $F(t)$ the kinetic frictional force between a single roller and the plank. (It is not necessary to assume that this force is constant in time.) During a short interval Δt the change in the angular momentum of the roller is

$$I\Delta \omega = rF(t)\Delta t. \tag{2}$$

These changes can be summed to give an equation containing the final angular velocity of the roller

$$r \sum F(t)\,\Delta t = I\omega_{max} = I\,\frac{v_{max}}{r}. \tag{3}$$

On the other hand, during time interval Δt the work done against friction (heat gain) ΔQ is the product of the kinetic frictional force and the *relative* displacement of the surfaces involved:

$$\Delta Q = F(t)\,[v_{max} - r\omega(t)]\,\Delta t.$$

From (2) and (3), the total dissipated energy is

$$\begin{aligned}
Q &= \sum F(t)\,[v_{max} - r\omega(t)]\,\Delta t \\
&= r\omega_{max} \sum F(t)\,\Delta t - I \sum \omega\,\Delta\omega \\
&= I\omega_{max}^2 - I\frac{\omega_{max}^2}{2} = I\frac{\omega_{max}^2}{2}.
\end{aligned}$$

In the final line we have used the fact that $\omega\,\Delta\omega = \frac{1}{2}\Delta\left(\omega^2\right)$. This result shows that the dissipated heat is equal in magnitude to the kinetic energy acquired by the rollers. It is remarkable that the result depends on neither the magnitude of the frictional force nor its time-dependence. The *correct* energy balance is not equation (1), but

$$MgL\sin\alpha = \frac{L}{d}\,\frac{1}{4}mv_{max}^2 + Q = 2\,\frac{L}{d}\,\frac{1}{4}mv_{max}^2,$$

which shows that the terminal velocity is

$$v_{max} = \sqrt{\frac{2dMg\sin\alpha}{m}}.$$

S96 Treating the problem as two-dimensional, choose a point P on the surface of the table and examine the angular momentum of the ball about this point. The line of action of the frictional force passes through this point, and so there is no frictional torque about P. The gravitational force and the supporting reaction of the table nullify each other. No other force acts on the ball, which therefore has constant angular momentum about the chosen point. As the ball is initially at rest, the value of that angular momentum is zero.

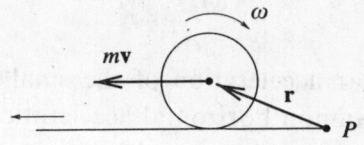

When the tablecloth is pulled out from under it, the ball starts sliding and rolling. Following the notation in the figure, its angular momentum **J** can be written as the sum of two terms:

$$\mathbf{J} = I\omega + \mathbf{r} \times (m\mathbf{v}).$$

Here I is the moment of inertia and m the mass of the ball. The first term in the expression is the internal spin, and the second the orbital angular momentum due to the linear motion of the centre of mass. Taking into account the directions of the vectors involved, the magnitude of the angular momentum can be written as $J = I\omega + mvR$, where R is the radius of the ball.

It is easy to see that, when any ball is rolling without slipping, the sign of its orbital angular momentum has to be the same as that of its spin. On the other hand, here the sum of the two has to be zero at all times. These two conditions can only be fulfilled at the same time if the body has stopped. The *reader* can check this experimentally.

The final state depends neither on the size of the frictional force, nor on how the tablecloth is pulled out. (It can be pulled out evenly, with a uniform acceleration, or by means of several sudden movements.) However, it is important that air resistance and rolling resistance are negligible since their effects can change the angular momentum about P.

S97   Taking the Earth's actual direction of rotation (from west to east) as positive, the angular momentum of the traffic about the axis of rotation would increase if the change were made. This is because the traffic that is travelling eastward would move to a greater distance from the Earth's axis thus increasing its (positive) contribution to the total angular momentum; conversely the westward-bound traffic would reduce its negative contribution. Assuming equal amounts of east–west and west–east traffic, the moment of inertia of the system is unchanged and, since the total angular momentum of the system cannot change, the Earth's rate of rotation must decrease. The length of the day would therefore increase – but you would hardly notice it!

> *Note.* One can also arrive at the same conclusion in a different way. In Great Britain there are a lot of traffic roundabouts. Any change in the direction in which these were negotiated would cause a change in the angular momentum of the traffic, which in turn would cause a small change in the rotation of the Earth. The whole traffic system can be considered as a series of many roundabouts.

S98   Let the angular acceleration of the smaller ball be α_1, that of the larger one α_2, their common horizontal acceleration a_1 and the acceleration of the cart a_2. As the balls are rolling without slipping, we have

$$R\alpha_2 = a_2 - a_1 \qquad \text{and} \qquad R\alpha_2 = r\alpha_1,$$

and, because $R = 2r$,

$$\alpha_1 = 2\alpha_2 = \frac{a_2 - a_1}{r}.$$

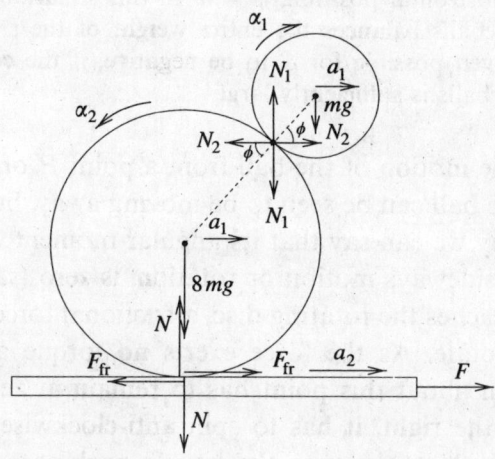

The moment of inertia of the smaller ball is $\frac{2}{5}mr^2$, while that of the larger one with the same density is $\frac{2}{5} \times 8m \times (2r)^2 = \frac{64}{5}mr^2$. Using the notation of the figure, we can write the following dynamical equations of motion:

$$F - F_{\text{fr}} = Ma_2,$$

$$8mg + N_1 - N = 0, \qquad F_{\text{fr}} - N_2 = 8ma_1,$$

$$mg - N_1 = 0, \qquad N_2 = ma_1,$$

$$N_1 r \cos \phi - N_2 r \sin \phi = \frac{2}{5}mr^2\alpha_1,$$

$$2rF_{\text{fr}} + 2rN_2r \sin \phi - 2rN_1r \cos \phi = \frac{64}{5}mr^2\alpha_2.$$

From these equations we can express the force F as

$$F = \left(9m + \frac{7}{2}M\right) \frac{\cos \phi}{1 + \sin \phi} \, g \approx 79 \text{ N}.$$

The acceleration of the balls relative to the cart is

$$\Delta a = a_2 - a_1 = \frac{5}{2} \frac{\cos \phi}{1 + \sin \phi} \, g.$$

At the time t when the balls fall from the cart, the distance they have moved relative to the cart is $L/2$. As their initial velocities are zero,

$$t = \sqrt{\frac{L}{\Delta a}} = 0.55 \text{ s}.$$

Note. It is interesting that this stunt can also be performed with the smaller ball in the horizontal position, $\phi = 0$. In this situation the frictional force between the balls balances the entire weight of the smaller ball. What is more, it is even possible for ϕ to be negative, if the coefficient of friction between the balls is sufficiently large!

S99   Observe the motion of the ball from a point P on its original direction of motion. The ball can be seen to be moving away, but has no sideways velocity. This is why we can say that its angular momentum about P, which would result from sideways motion or rotation, is zero (*see also* P96).

When the ball reaches the rotating disc, a frictional force causes it to move sideways and to rotate. As the force exerts no torque about P, the total angular momentum about this point has to remain at zero at all times. (If the ball moves to the right, it has to spin anti-clockwise about its vertical axis, and conversely if it moves to the left. In each case it spins with such angular velocity as is needed to maintain zero total angular momentum.)

Similar effects occur when the ball leaves the disc and rolls onto the stationary table. Here, however, if the ball moved to the right, it would spin clockwise, in accordance with the condition for rolling without slipping. This is impossible, as the angular momenta due to these two types of motion would then have the same sign and their resultant could not be zero. The only solution is to exclude any final motion of the ball perpendicular to the original direction of its motion, i.e. to allow that, when it returns to the stationary table, the ball rolls on without having changed its overall direction.

Similar reasoning shows that its speed after leaving the disc cannot be different from its original speed. This is a consequence of the conservation of angular momentum about an axis perpendicular to the direction of motion.

Thus, the ball leaves the table at the same speed, and with the same momentum vector, as it had originally. Further, its total kinetic energy is unchanged. This last fact is especially strange, as both when the ball arrives at the disc and when it leaves it, friction does work on the ball and changes its kinetic energy. However, the algebraic sum of the work done on the ball is zero; this is not a consequence of the conservation of either momentum or energy, but of the conservation of angular momentum.

Note. Rather more complicated calculations (using the equations of translational and rotational motions) show that (viewed from the frame of reference of the table) the ball follows a uniform circular motion on a steadily rotating disc. This circle, however, is not centred on the axis of the disc, and the motion's angular velocity is different from that of the disc, being $\frac{2}{7}$ times smaller. If the disc rotates steadily, the ball ultimately continues along the extrapolation of its original track, but this is not true if the disc does not rotate uniformly. In the latter case, the magnitude of the ball's momentum is unchanged, but its track deviates sideways as a result of its encounter with the disc.

S100   Let the tension in the ring be T. Its resolved component acting along the radius towards the centre of rotation is $2T\sin(\Delta\theta/2) \approx T\Delta\theta$ and this must balance the centripetal force of $R\Delta\theta A\rho R\omega^2$ (*see figure*).

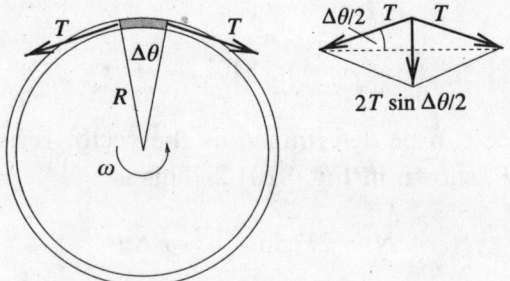

It follows that the longitudinal stress in the ring, T/A, is $\rho R^2\omega^2$; the strain ε is E^{-1} times this. Finally, the increase in circumference, given by $2\pi R\varepsilon$, is $2\pi\rho R^3\omega^2/E$.

S101   When one end of the thread is pulled by a force F_0, let the maximum force with which the other end can be pulled without the thread slipping on the cylinder be $F_{\max}$. Specify a general point of the thread in contact with the cylinder by the angle α, which the radius of the cylinder at that point makes with a fixed radius. When the thread wound onto the cylinder is tightened, it exerts a normal force on the cylinder resulting in a frictional force which opposes any relative motion of the thread and the cylinder. The tension in the thread increases as α increases, but the excess tension at one end of a piece of the thread is balanced by the frictional force acting on that piece.

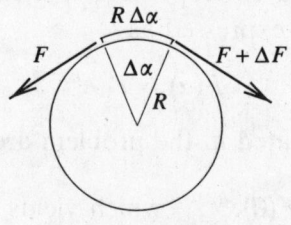

Fig. S101.1

Consider a small length of thread that subtends an angle $\Delta\alpha$ at the centre of the cylinder. If, as shown in Fig. S101.1, the tension at one end of the small piece is F whilst it is $F + \Delta F$ at the other, then the excess force ΔF is balanced by a frictional force, which can be calculated as

$$\Delta F = \mu N, \tag{1}$$

where N is the force exerted by the thread normal to the surface of the cylinder and μ is the required coefficient of friction.

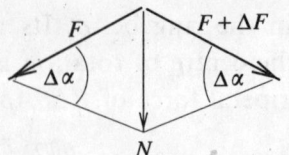

Fig. S101.2

The normal force can be determined as the vector resultant of the forces F and $F + \Delta F \approx F$, shown in Fig. S101.2. This is

$$N = 2F \sin\frac{\Delta\alpha}{2} \approx F\Delta\alpha. \tag{2}$$

Substituting this into equation (1) shows the relationship between F and the angle α to be

$$\Delta F(\alpha) = \mu F(\alpha)\,\Delta\alpha.$$

This relationship is formally similar to the equation governing radioactive decay,

$$\Delta m(t) = -\lambda\, m(t)\,\Delta t,$$

where $m(t)$ is the mass of radioactive material, t the elapsed time, and λ the decay constant. As is well known, the mass of radioactive material decreases exponentially with time, i.e.

$$m(t) = m_0\, e^{-\lambda t}.$$

Thus, using the established correspondence, with $-\lambda$ replaced by μ, the law of 'thread friction' can be expressed as

$$F(\alpha) = F_0\, e^{\mu\alpha}. \tag{3}$$

Both of the inequalities stated in the problem are equivalent to

$$2F(0) = F(\pi) = F(0)e^{\mu\pi}, \quad \text{which yields } \mu = \frac{1}{\pi}\ln 2 \approx 0.22.$$

Note. The force exerted on the thread increases exponentially with the angle α. The ratio of the forces at the two ends of the thread can reach a large order of magnitude after only a few turns. Climbers make use of this interesting fact when they anchor the ropes that prevent them from falling. Sailors use the same technique to stop large boats with their bare hands!

S102 Jenny considers a homogeneous ring of radius R and linear density ρ, rotating with constant angular velocity ω about an axis perpendicular to its plane and passing through its centre. The centripetal acceleration of points on the ring is $R\omega^2$, and so unit length of it experiences a centripetal force of $p = \rho R\omega^2$. If a cylindrical container of height 1 m were surrounded by a gas at pressure p then the force exerted by the gas on the wall of the cylinder would be exactly the force required to sustain the rotation (see Fig. S102.1).

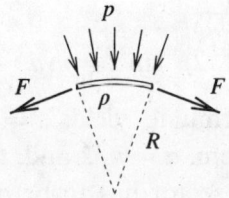

Fig. S102.1

In reality, the elements of the ring are not kept in their circular orbit by some imaginary external pressure, but by the ring's own internal tension, whose magnitude is

$$F = \rho R^2 \omega^2.$$

This can be proved by referring to P100 or by examining, for example, a 1-m length of a container with a semicircular base, surrounded by a medium at pressure p (*see* Fig. S102.2). A force of $2Rp$ acts on the rectangle of area $2R$, and has to be balanced by a force of magnitude $2F$ acting tangentially within the wall.

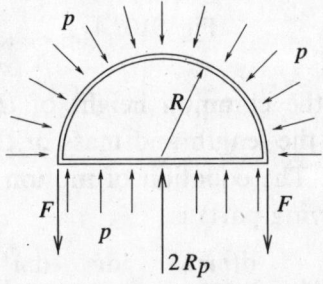

Fig. S102.2

Now examine an arc of the ring which subtends an angle at its centre of 2α, and see how Newton's second law of motion is fulfilled, continues Jenny. The resultant of the forces acting on a body of mass $m = 2R\rho\alpha$ is $2F\sin\alpha$, and the acceleration of the centre of mass is $a = s\omega^2$, where s denotes the distance between the centre of mass of the arc and the centre of the ring (*see* Fig. S102.3).

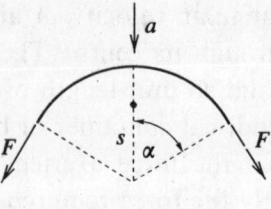

Fig. S102.3

According to Newton's law,

$$2F\sin\alpha = ma,$$

which, using the previous formulae, yields $s = (R\sin\alpha)/\alpha$. For a semicircle, as in Charlie's original problem, $\alpha = \pi/2$ and, therefore, $s = 2R/\pi$.

If the centre of mass of a sector has to be determined, it can be divided into thin arcs and each one replaced by a point-like body of appropriate mass positioned at the centre of mass of the arc. The same procedure could be adopted to find the centre of mass of a triangle made up of thin stripes. This implies that the centre of mass of a sector has to be in exactly the same place as that of a symmetrical triangle of height $s_{\max} = (R\sin\alpha)/\alpha$ (see Fig. S102.4), i.e. at a distance of $\frac{2}{3}(R\sin\alpha)/\alpha$ from its vertex. With this deduction Jenny concludes her display of extraordinary logic.

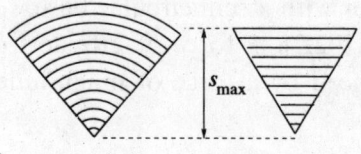

Fig. S102.4

S103 Let us denote the common height of the table and total chain length by $L(= 1 \text{ m})$ and the length and mass of the vertically moving part of the chain by x and m. The equation of motion (taking into account the changing mass of the moving part) is

$$mg - \frac{\mathrm{d}(mv)}{\mathrm{d}t} = m\frac{\mathrm{d}v}{\mathrm{d}t} + \frac{\mathrm{d}m}{\mathrm{d}t}v,$$

which on rearrangement gives

$$m\frac{dv}{dt} = mg - \frac{dm}{dt}v.$$

The left-hand side is the product of the instantaneous mass and acceleration a, whilst the right-hand side can be converted and simplified using $dm = (m/x)dx$ and $dx/dt = v$ to yield

$$a = g - \frac{v^2}{x}.$$

This result shows that the acceleration of the chain is less than g. The second term on the right-hand side can be simplified further, since $v^2/x = 2f$ in the case of rectilinear motion with constant acceleration f and zero initial velocity. This means that in the current problem the acceleration of the chain is constant and satisfies

$$a = g - 2a, \qquad \text{yielding } a = \frac{g}{3}.$$

As the chain runs down from the table during a time of t_1, its first link falls a distance L with acceleration $g/3$. Consequently,

$$t_1 = \sqrt{\frac{2L}{a}} = \sqrt{\frac{6L}{g}} = 0.78 \text{ s}.$$

When the lower end of the chain reaches the ground, the whole chain is vertical and its velocity is

$$v_1 = at_1 = \frac{g}{3}\sqrt{\frac{6L}{g}} = \sqrt{\frac{2Lg}{3}} = 2.56 \text{ m s}^{-1}.$$

From this moment on, the chain goes into free fall. Its last link has an initial velocity v_1, accelerates with g, and covers a distance L in time t_2. Thus

$$L = v_1t_2 + \frac{1}{2}gt_2^2.$$

From this equation we obtain

$$t_2 = \sqrt{\frac{2L}{3g}} = \frac{t_1}{3} = 0.26 \text{ s}.$$

So the final link of the chain reaches the floor at a time

$$t_1 + t_2 = \frac{4}{3}t_1 = 1.04 \text{ s}$$

after the start of the process.

Note. (i) Attempting to apply the law of conservation of energy leads to *false* results. For example, setting

$$\frac{1}{2}mv_1^2 = mg\frac{L}{2},$$

where $L/2$ is the loss of height of the centre of mass of the chain, and then substituting $v_1 = \sqrt{2Lg/3}$ leads to the contradictory result

$$mg\frac{L}{2} = \frac{1}{2}mv_1^2 = \frac{1}{2}m\left(\sqrt{\frac{2}{3}Lg}\right)^2 = mg\frac{L}{3}.$$

It would appear that one-third of the energy has disappeared. This is, in fact, accounted for by the energy dissipated in the series of inelastic collisions occurring when the chain jerks the successive links into motion.

(ii) The problem can also be solved in a different way. Let M be the total mass of the chain. When the hanging part of the chain of mass $m = (M/L)x$ causes the next piece, of mass $(M/L)\Delta x = (M/L)v\Delta t$, to move, it accelerates the piece from rest to a velocity v in a time interval of Δt. This acceleration needs a force of

$$\frac{[(M/L)v\Delta t]\,v}{\Delta t} = \frac{M}{L}v^2.$$

The corresponding reaction decelerates the hanging part of the chain, so we can write

$$ma = mg - \frac{M}{L}v^2.$$

Inserting $m = (M/L)x$ into this equation, we recover the earlier solution.

(iii) Assuming that the chain consists of n links with an otherwise unconstrained separation of $\varepsilon = L/n$ between links, leads to the correct answer in the limit $n \to \infty$.

(iv) It is possible, using calculus techniques, to solve the (non-linear) differential equation

$$a = g - \frac{v^2}{x},$$

subject to the initial conditions $v = 0$, $x = x_0$ ($x_0 \ll L$) at $t = 0$. The solution

$$v(x) = \sqrt{\frac{2}{3}gx\left[1 - \left(\frac{x_0}{x}\right)^3\right]}, \qquad t = \int_{x_0}^{x}\frac{dx}{v(x)}$$

approaches our more heuristic result in the limit $x_0 \to 0$.

S104 It will be shown that a chain (flexible rope), moving at a uniform speed along a closed curve of arbitrary shape, continues moving in the same way even if no constraints (e.g. pulleys, cylinders, etc.) are placed on it.

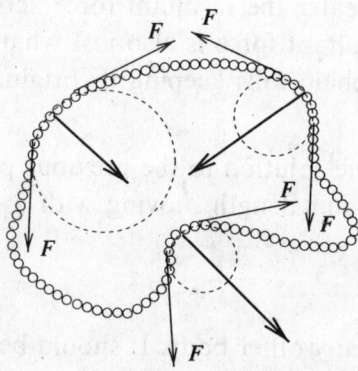

Fig. S104.1

Consider the chain (shown in Fig. S104.1) moving steadily along some closed, winding space-curve at speed v. The force stretching the chain has to have the same magnitude, F, everywhere, as the tangential acceleration of its links is zero. If the radius of curvature of the chain, of mass per unit length ρ, is R at some point (R can vary from place to place), then the mass of a piece of length $R\Delta\alpha$ is $\Delta m = \rho R\Delta\alpha$, whilst its acceleration is v^2/R, as shown in Fig. S104.2. Its equation of motion is

$$\rho R\Delta\alpha \, \frac{v^2}{R} = F\Delta\alpha,$$

which leads to the relation $F = \rho v^2$.

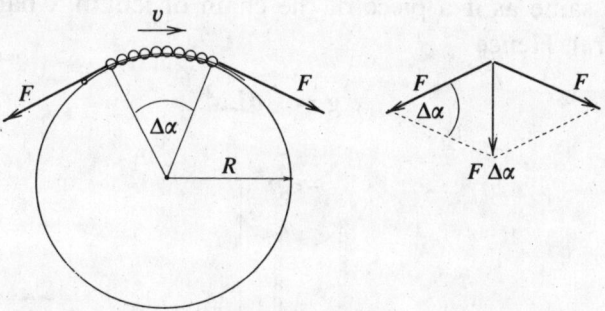

Fig. S104.2

Notice that R does *not* occur in this equation, i.e. F is independent of the radius of curvature. The resultant of the tangential forces, F, is just the right force to make the chain curve as it does at the given place. If the chain is straight, the resultant force acting on a small piece of it is zero. The more curved the chain, the greater the resultant force acting on each small piece. The direction of the resultant force is also just what is needed.

This means that the chain falls keeping its original shape and speed, just as Frank guessed.

S105   As shown in the solution to the previous problem, a flexible chain, or rope, of mass ρ per unit length moving with a speed v has an internal tension of

$$F = \rho v^2$$

if it is not contact with any other body. It should be noted that this result is independent of the radius of curvature R of the arc formed; it also applies to chains or ropes moving in a straight line when R can formally be taken as infinite. In the present case, the chain becomes detached from the pulley, as a result of its accelerating motion, when this freedom condition is satisfied.

Denote the displacement of the chain by x and the acceleration of the right-hand side of the chain by a. The equations of motion for the two sides are

$$F - \rho \left(\frac{L}{2} - x\right) g = \rho \left(\frac{L}{2} - x\right) a,$$

$$\rho \left(\frac{L}{2} + x\right) g - F = \rho \left(\frac{L}{2} + x\right) a.$$

The speed of the chain for any x can be determined, without using integration, from the conservation of energy. The decrease in potential energy relative to that of the original situation (for which negligible speed is assumed) is the same as if a piece of the chain of length x had been lowered by x (*see figure*). Hence

$$\rho x^2 g = \frac{1}{2}\rho L v^2.$$

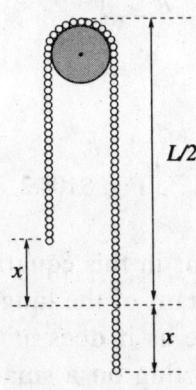

The displacement and speed when the chain leaves the pulley can be calculated by eliminating F and a from the above equations:

$$x = \frac{1}{2\sqrt{2}}L \approx 0.35\,L. \qquad v = \frac{\sqrt{Lg}}{2}.$$

Thus, the chain becomes detached from the pulley when 15 per cent of its length is still moving upwards and its speed is less than one would obtain by naïvely substituting $x = L/2$ into the conservation of energy equation. (This false value would be $\sqrt{Lg/2} \approx 0.71\sqrt{Lg}$.)

> *Note.* The subsequent motion of the chain is also interesting. The loop, once detached from the pulley, starts moving upwards with increasing speed. The speed of this piece of chain, of decreasing mass, tends, in principle, to infinity. In reality, the finite size of the links of the chain and air resistance impose an upper limit on the speed. The kinetic energy of the piece of chain moving upwards remains finite – despite its rapidly increasing speed – because the decrease in the mass of the relevant part of the chain is more rapid than the increase in its speed. The same phenomenon can be observed when a whip cracks; a section (of decreasing length) of the whip moves at an ever increasing speed, and when it reaches the speed of sound, it causes a sharp supersonic bang.

S106 (i) Examine the motion of the loop in the frame of reference, which moves with the loop at speed c. In this system, the pieces of a circular loop of radius R rotate uniformly. The acceleration of a piece of the rope, which subtends central angle $\Delta\alpha$ and has mass $\rho R\,\Delta\alpha$, is c^2/R, whilst the net force due to the tension in the rope is $F\,\Delta\alpha$ (see S105). The Newtonian law of motion yields

$$F\,\Delta\alpha = \rho R\,\Delta\alpha\,\frac{c^2}{R},$$

i.e. the 'loop-wave' moves with speed $c = \sqrt{F/\rho}$, – identical to the speed at which small transverse waves would propagate along the same rope.

(ii) The angular frequency of the rolling loop of radius R is $\omega = c/R$, and the loop has mass, $m = 2\pi R\rho$. The energy carried by the loop can be expressed in terms of these quantities as

$$E = E_{\text{transl}} + E_{\text{rot}} = \frac{1}{2}(2\pi R\rho)(c^2 + R^2\,\omega^2),$$

which can be written in the form

$$E(\omega) = 2\pi Fc\,\frac{1}{\omega} = K\,\frac{1}{\omega}.$$

The quantity $K = 2\pi Fc$ is a constant, characteristic of the rope and the tension in it. Of course, several (n) loops can be simultaneously excited and their total energy is

$$E_n(\omega) = nK\frac{1}{\omega} \quad (n = 0, 1, 2, 3, \ldots).$$

The momentum of the loop(s) can be calculated in a similar way,

$$P_n(\omega) = n(2\pi R\rho)c = nK\frac{1}{c\,\omega} \quad (n = 0, 1, 2, 3, \ldots),$$

as can the angular momentum

$$J_n(\omega) = \pm n\,(2\pi R\rho)\,Rc = \pm nK\frac{1}{\omega^2} \quad (n = 0, 1, 2, 3, \ldots).$$

In this latter formula, the two signs correspond to loops moving 'above' and 'below' the rope, i.e. to the two possible 'polarisations'.

It can be seen that – provided only circular excitations of a single frequency are allowed – the energy, momentum and angular momentum can only assume discrete values, those that can be written as the product of an integer and a basic 'quantum'. The following relationships are valid between these (frequency-dependent) 'quanta':

$$E(\omega) = cP(\omega) = \omega J(\omega).$$

It is not difficult to recognise that the same relationships are valid for photons

$$E_{\text{photon}} = \hbar\omega; \quad P_{\text{photon}} = \hbar\omega/c; \quad J_{\text{photon}} = \pm\hbar.$$

Naturally, this formal analogy must not be taken too seriously, e.g. by thinking of a photon as equivalent to a loop. However, the similarity can be used to show that, even in classical physics, there are objects more complicated than a point mass which are easy to understand, but which can still be excited to many discrete energy, momentum and angular momentum levels.

S107 A volume of sand of mass $\Delta m = 50$ kg reaches a speed of $v = 1$ m s^{-1} in time $\Delta t = 1$ s. The change in its horizontal momentum is therefore $\Delta p = \Delta m v = 50$ kg m s^{-1}. This means that a force

$$F = \frac{\Delta p}{\Delta t} = \frac{v\,\Delta m}{\Delta t} = 50 \text{ N}$$

accelerates the sand. The work done by the engine – taking only the acceleration of the sand into account – is 50 J s^{-1}, i.e. its power output is 50 W.

The sand loses its vertical momentum when it lands on the conveyor belt. (It hits the belt vertically with a force greater than its weight.) The sand on the belt is slowed down vertically and then accelerated horizontally; the belt's kinetic energy is increased. The kinetic energy of the sand increases by $\Delta m\, v^2/2 = 25$ J in $\Delta t = 1$ s. This means that one-half of the power of the engine (25 W) is converted into kinetic energy of the sand; the rest is the work done against friction and converted into heat.

> *Note.* The average speed of the sand during the acceleration is $v/2$. Therefore the power of the frictional force ($F = 50$ N) is $Fv/2 = 25$ W. The belt experiences a force $-F$, the power of which is $-Fv = -50$ W. Thus, exactly one-half of the power of the engine is used to accelerate the sand.

S108 (i) First, we determine the speed of the roll as a function of the distance it has covered. The mass of the moving part of the hose after travelling a distance, x, is $m(x) = M(1 - x/L)$. Its speed $v(x)$ can be determined using the conservation of energy:

$$\frac{1}{2}Mv_0^2 + \frac{1}{2}\left(\frac{1}{2}MR^2\right)\left(\frac{v_0}{R}\right)^2 = \frac{1}{2}mv^2 + \frac{1}{2}\left(\frac{1}{2}mr^2\right)\left(\frac{v}{r}\right)^2.$$

The change in potential energy and the small vertical speed acquired as a result of the decrease of the radius of the roll have been neglected. Using the known variation of the mass with distance, the velocity v is found to be given by

$$v(x) = \frac{v_0}{\sqrt{1 - x/L}}.$$

As x increases, so does the velocity, i.e. the roll accelerates as it unrolls.

The total time taken to unroll can be obtained by integrating the reciprocal of the function $v(x) = dx/dt$:

$$T = \int_0^T dt = \int_0^L \frac{dx}{v(x)} = \frac{1}{v_0}\int_0^L \sqrt{1 - x/L}\, dx = \frac{L}{v_0}\int_0^1 \sqrt{1 - u}\, du = \frac{2}{3}\frac{L}{v_0}.$$

Since the hose accelerates, the time taken to unroll is obviously shorter than if the hose had unrolled with uniform velocity. In fact, it is two-thirds of that figure.

(ii) The system consisting of a roll of decreasing mass and increasing speed, and of a motionless horizontal part of increasing length can obviously not be considered as a point mass! Therefore the basic law of dynamics cannot be applied to it in the simple form $F = ma$, but must be used in the more general form

$$\sum F_{\text{external}} = \frac{d}{dt}p_{\text{total}},$$

where p_{total} is the momentum of the system as a whole.

The total momentum of the moving roll (and of the whole system) is

$$p(x) = m(x)\,v(x) = M(1 - x/L)\frac{v_0}{\sqrt{1 - x/L}} = Mv_0\sqrt{1 - x/L}.$$

Clearly, as x increases, $p(x)$ decreases – reflecting the fact that the mass of the piece in motion decreases faster than the rate at which its speed increases. The direction of the resultant force $K(x)$ acting on the system is, therefore, opposite to the direction of the motion, with

$$K(x) = \frac{dp}{dt} = \frac{dp(x)}{dx}v(x) = -\frac{Mv_0^2}{L}\frac{1}{2(1 - x/L)}.$$

S109   The gravitational field due to a thin spherical shell of uniform mass distribution is zero inside the shell. Outside it is as if the total mass of the shell were concentrated at its centre. At a distance of 100 km below the surface of the Earth, two factors affect the gravitational field. On the one hand, the mass of the part of the Earth still 'underneath' is smaller than the total mass of the Earth, meaning that the gravitational acceleration will be reduced. On the other hand, the centre of the Earth is closer, which will tend to increase g. Which effect is the stronger?

The shell of thickness 100 km corresponds to 4.6 per cent of the total volume of the Earth, which has a radius of 6400 km, but its mass is only 2.5 per cent of the total mass of the Earth. Gravitational acceleration can be calculated as $g = GM/r^2$, where M is the mass inside the radius r. At a depth of 100 km below the surface, the effective mass of the Earth (without its crust) is $M' = 0.975\,M$, with radius $r' = (6300/6400)\,r = 0.984\,r$. Substituting this data into the above equation, we find that 100 km below the surface of the Earth, the gravitational acceleration is 0.7 per cent *greater* than at the surface!

More generally, it can be proved that g increases as the centre is approached if the density of the crust is not **greater** than two-thirds of the average density.

S110   First consider the trial bore, which is of negligible volume compared with that of the whole asteroid. Let the density of the asteroid be ρ and its radius R. The gravitational acceleration at radius r is the same as if only the sphere of radius r below it were present:

$$g(r) = G\frac{m(r)}{r^2} = G\frac{(4/3)\pi r^3 \rho}{r^2} = \frac{4\pi G\rho}{3}r.$$

Thus the gravitational acceleration is directly proportional to the distance from the centre of the asteroid and always points towards it.

This means that the first unfortunate little green man approached his fate whilst executing a harmonic oscillation of amplitude R, and reached it after the first quarter-period of the oscillation. The coefficient of r in the above expression corresponds to ω^2, i.e. to the square of the angular frequency. Therefore the duration of the fall was

$$T_1 = \frac{T}{4} = \frac{1}{4}\sqrt{\frac{3\pi}{G\rho}}.$$

The speed at which he collided with the bottom of the hole can be obtained as the product of his amplitude and angular frequency,

$$v_1 = R\omega = 2R\sqrt{\frac{\pi G\rho}{3}}.$$

By the time of the second accident, the little green people had already excavated one-eighth of the material in the asteroid. The gravitational field in the spherical cavity, extending from the surface of the planet to its centre, has to be determined. This can be done using a 'cunning' application of the principle of superposition: imagine the cavity to be filled with a mixture of 'normal' titanium and 'negative density' titanium.

The gravitational fields of the complete asteroid and of the sphere of 'negative titanium' have to be added. The vector $\mathbf{r}$ pointing to an arbitrary point P in the cavity, the vector $\mathbf{c}$ pointing from the centre of the cavity to the centre of the asteroid and the vector $\mathbf{r} + \mathbf{c}$ pointing from the centre of the cavity to point P are shown in Fig. S110.1.

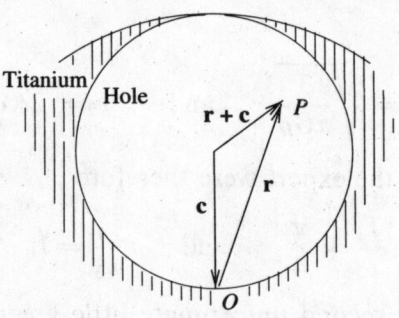

Fig. S110.1

The gravitational accelerations (of the homogeneous asteroid, of the cavity and of their resultant) are proportional to the position vectors, the coefficient of proportionality being the constant previously denoted by $-\omega^2$. The 'lack of matter' in the cavity is represented by a negative density.

The resultant acceleration (*see* Fig. S110.2) is

$$\mathbf{g} = -\omega^2\mathbf{r} + \omega^2(\mathbf{r} + \mathbf{c}) = \omega^2\mathbf{c}.$$

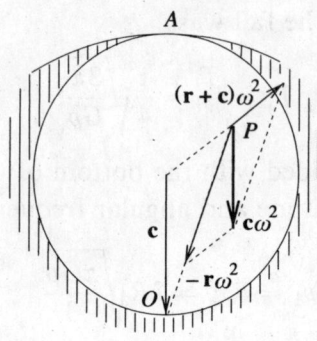

Fig. S110.2

This gravitational acceleration is a constant vector, regardless of the position of point P inside the cavity. This means that there is a homogeneous gravitational field in the cavity, with a magnitude $-\omega^2 R/2$, the gravitational acceleration in the middle of the trial bore (i.e. at the centre of the cavity).

> *Note.* It can similarly be shown that there is a homogeneous field in any spherical cavity; it has a magnitude equal to the gravitational acceleration at the position inside the solid sphere on which the cavity is to be centred.

The duration T_2 of the fall and the speed of impact v_2 of the second unfortunate little green man can be calculated by applying the equations for *uniform* acceleration:

$$T_2 = \frac{2}{\omega} = \sqrt{\frac{3}{\pi G \rho}} \quad \text{and} \quad v_2 = 2R\sqrt{\frac{\pi G \rho}{3}}.$$

The ratios reported by the expert were therefore

$$\frac{T_1}{T_2} = \frac{\pi}{4} \quad \text{and} \quad \frac{v_1}{v_2} = 1.$$

It can be seen that the second unfortunate little green man took a slightly longer time to fall than the first one, but that they both hit the centre of the asteroid with the same speed.

The fact that the speeds are the same – and, by implication, so are their kinetic energies – is not a coincidence! In the first case, because the gravitational potential is lower at the centre of the asteroid than at its surface, the first unfortunate man is accelerated and gains the corresponding amount of kinetic energy. The second case is different in that the gravitational potential of the 'negative density' cavity has to be taken into account as well. However, the man fell from one edge of the cavity to the other, and so his gravitational potential with respect to the cavity did not change. Applying the principle of superposition to the gravitational potentials gives the final result, namely that the total change in the potentials is the same in both cases, and so the speeds acquired by the two victims are the same.

S111 Imagine the hemisphere to be divided into many concentric hemispherical shells of identical thickness. What is the force exerted by these shells on a probe of unit mass sitting in the centre of the sphere? Since the mass of a shell is *directly* proportional to the square of its radius and the force exerted by a given-mass shell on the probe is *inversely* proportional to the square of its radius, the gravitational accelerations at the centre due to the different hemispherical shells are all equal (*see* Fig. S111.1).

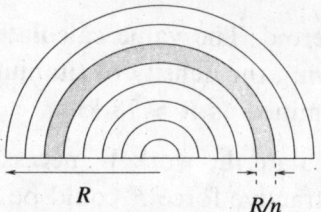

R R/n

Fig. S111.1

If there are n shells, then the mass of the outermost shell is $2\pi R^2 (R/n)\rho$, where R is the radius of the asteroid and ρ is its density. The total gravitational field due to the hemisphere is n times that due to its outermost shell, i.e. it is the same as if the surface of the hemisphere had mass $M = 2\pi R^3 \rho$. This mass is, in fact, three times the actual mass of the hemisphere.

What force is exerted by this hemispherical shell of mass M on the probe? Considering only its magnitude, the force is the same as that exerted by the unit-mass probe on the shell, i.e. $p = G(M/2\pi R^2)(1/R^2) = G\rho/R$ per unit surface area. To integrate this effect over the whole shell, the situation can be compared with that of finding the force exerted by a liquid at pressure p on a similar hemispherical shell (*see* Fig. S111.2).

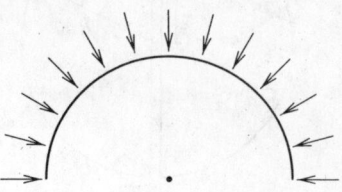

Fig. S111.2

Since the resultant force of a liquid acting on a complete hemisphere would be zero, the total force acting on its curved surface would be the same in magnitude as the force acting on its plane surface. Thus $g = p\pi R^2 = G\rho R\pi$ (*see* Fig. S111.3).

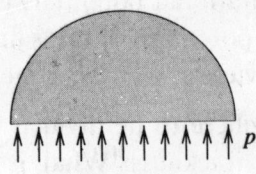

Fig. S111.3

Now consider the initial gravitational acceleration,

$$g_0 = G\frac{4\pi R^3}{3}\rho\frac{1}{R^2},$$

at the surface of the asteroid. The value calculated above is therefore $g = \frac{3}{4}g_0 = 7.36$ cm s^{-2}. Knowing the density of titanium, the (original) radius of the asteroid can be determined as $R \approx 78$ km.

S112   If we could calculate the work W necessary to pull the two halves apart by $d = 1$ m, the attractive force F could be found from $W = Fd$. But how can W be determined? Certainly not directly, as the product force × displacement, since that would lead us back to the original problem. Some other method has to be found.

The gravitational potential energy of the system increases when the hemispheres are pulled apart. W has the same value as this increase. The increase can be calculated by determining the work done by the little green people when carrying to the surface the titanium originally in the disc of radius R and thickness d. The extracted metal can be considered as having been carried to the surface by means of frictionless devices and then spread evenly over the asteroid's surface.

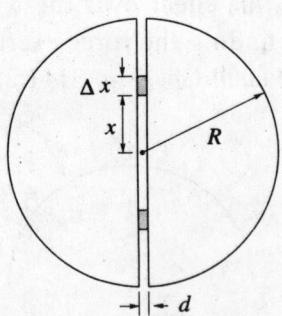

If g denotes the gravitational acceleration at the surface of the asteroid of mass M, then the gravitational acceleration at a distance x from the centre is $g(x) = xg/R$. The density of the asteroid, and hence of titanium, is $\rho = 3M/(4\pi R^3)$. Consider the metal, of volume $\Delta V = d\,2\pi x\,\Delta x$ and mass $\Delta m = \rho\Delta V$, which was originally situated between radii x and $x + \Delta x$ (see figure). The metal has all to be brought up from the same depth and can, therefore, be considered together when calculating the work done. A force, $\Delta mgx/R$ acts on this metal at its original position, whilst at the surface, the corresponding force is Δmg. Since the gravitational field increases uniformly when moving from inside the asteroid to its surface, the arithmetic mean of the initial and final forces can be used. The total displacement is $R - x$, and the work done is therefore

$$\Delta W = \Delta mg\frac{1 + (x/R)}{2}(R - x) = \Delta mg\frac{R^2 - x^2}{2R} = \frac{3}{4}\frac{Mgd}{R^4}(R^2 - x^2)x\,\Delta x.$$

The total work involved is the sum of that for bringing the titanium up from the different depths, explicitly

$$W = \sum\Delta W = \frac{3Mgd}{4R^4}\sum(R^2 - x^2)x\,\Delta x.$$

In the limit of layers of vanishingly small thickness, the above sum becomes an integral with value

$$\sum(R^2 - x^2)x\,\Delta x \Rightarrow \int_0^R (R^2 - x^2)x\,dx = \frac{R^4}{4}.$$

Therefore the total work done is $W = \frac{3}{16}Mgd$.

> Note. The same result can be obtained without using integral calculus. Introduce a new variable $u = x^2/R^2$, instead of x, $(0 \leq u \leq 1)$, and replace the summation over the layers at different depths by a summation over the terms corresponding to different values of u. Using the relationships $u = x^2/R^2$ and $u + \Delta u = (x + \Delta x)^2/R^2$ (and neglecting terms containing the square of the small quantity Δx), the work done can be expressed as
>
> $$W = \frac{3}{8}Mgd\sum(1 - u)\,\Delta u.$$
>
> This sum can be readily evaluated since $1 - u$ changes uniformly from 1 to 0, and can therefore be taken to be $\frac{1}{2}$ on average, whilst the sum of the terms in Δu is 1. In the end, the previous expression for the work done is obtained.

A 'mythical giant' does work $W = \frac{3}{16}Mgd$ when separating the hemispheres by d, and the force required to do this is therefore $F = 3Mg/16$. This is the force of attraction between the hemispheres and is the force the props should have borne. Knowing the radius of the asteroid and the density of titanium, this force can be determined numerically as

$$F = \frac{3}{16} M \frac{GM}{R^2} = \frac{3}{16} \left(\frac{4\pi R^3 \rho}{3} \right)^2 \frac{G}{R^2} = \frac{GR^4 \rho^2 \pi^2}{3} \approx 4.5 \times 10^{13} \text{ N.}$$

In order to get a feel for the order of magnitude of this force, we calculate the average pressure exerted on unit surface area as $p = F/(\pi R^2) = 1.4 \times 10^5$ Pa (i.e. one and a half times the atmospheric pressure on Earth). This is the weight, on Earth, of 14 tons of matter; sufficiently strong props should bear such a load.

S113   At the surface of the charged sphere, whether it consists of a single piece or two pieces close together, the electric field strength is

$$E = \frac{1}{4\pi\varepsilon_0} \frac{Q}{R^2}.$$

The electric charge per unit surface area is

$$\sigma = \frac{Q}{4\pi R^2}.$$

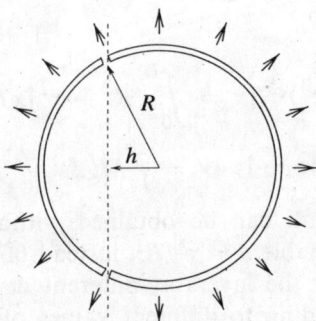

This electric field exerts a force $\Delta F = \frac{1}{2} E \Delta Q$ on the charge $\Delta Q = \sigma \Delta A$ which resides on a surface area ΔA, as illustrated in the figure. The reason for the factor of $\frac{1}{2}$ is that the electric field strength is E at the outer surface of the sphere and zero inside; its average value is therefore $E/2$.

The force per unit area exerted by the charges on the pieces of the sphere is therefore

$$\frac{\Delta F}{\Delta A} = \frac{Q^2}{32 \pi^2 \varepsilon_0 R^4} = p.$$

The required force can be compared with the force with which a liquid at pressure p would push apart the two pieces of the sphere. As this force is also the product of p and the cross-sectional area of the intersection of the plane and sphere, i.e. $p\pi(R^2 - h^2)$, it follows that the two parts of the sphere can be held together by a force

$$F = \frac{Q^2}{32\,\pi\varepsilon_0 R^4}(R^2 - h^2).$$

S114 Using the notation in the figure, the equilibrium condition for the first ball is

$$\frac{mg}{F} = \frac{\ell}{x},$$

where

$$F = kqQ/x^2$$

is the Coulomb force acting on the first ball and x is the distance between the balls carrying charges q and Q.

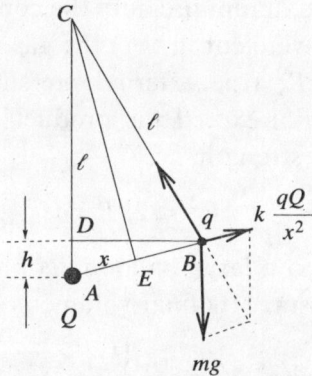

It is clear that the triangles ABD and CAE are similar, and that consequently

$$\frac{x}{2} : \ell = h : x.$$

From the three equations above we can calculate the separation of the charges and the electrostatic energy of the system:

$$x = k\frac{qQ}{2mgh} \qquad \text{and} \qquad E_{\text{electro}} = k\frac{qQ}{x} = 2mgh.$$

The work done is the sum of the changes in electrostatic and gravitational potential energy,

$$W = 2mgh + mgh = 3mgh.$$

It is perhaps surprising that the work done does not depend on either the magnitudes of the charges or the length of the thread.

S115   Hydrogen of pressure p is enclosed in a spherical container of radius R with walls of thickness d. Let ρ be the density of the wall material and σ its tensile strength, as illustrated in the figure.

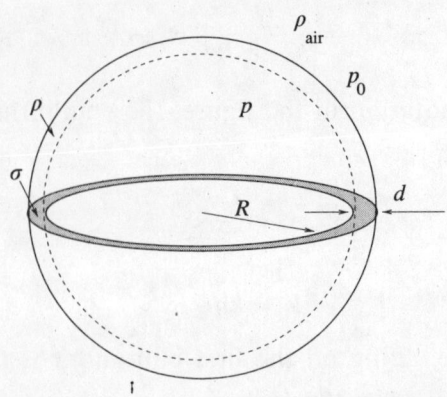

We first calculate the maximum pressure the container can sustain without bursting. If the container were cut in two, the gas would push the two pieces apart with a force of $p\pi R^2$ (the external pressure is negligible compared with p). This force must not exceed the product of the surface area of the cut, $2\pi R d$, and the tensile strength, i.e.

$$p\pi R^2 < 2\pi R \sigma d. \tag{1}$$

If the gas is released into a large balloon, its pressure decreases to p_0, the external atmospheric pressure. Its final volume is therefore

$$V = \frac{4\pi R^3}{3}\frac{p}{p_0}.$$

This volume of gas certainly cannot lift a load any greater than $V\rho_{\text{air}}g$ (where ρ_{air} is the density of air), as the weight of the balloon and the hydrogen have still to be subtracted from the upthrust.

The weight of the container is $4\pi R^2 d\rho g$. In accordance with the above reasoning, this cannot be greater than the force available to lift it, i.e.

$$4\pi R^2 d\rho g < \frac{4\pi R^3}{3}\frac{p}{p_0}\rho_{\text{air}}g. \tag{2}$$

Inequalities (1) and (2) give a relationship between the properties of the container material and the pressure and density of the external air,

$$\frac{\sigma}{\rho} > \frac{3p_0}{2\rho_{\text{air}}}. \tag{3}$$

It is interesting that neither the radius, nor the thickness of the walls of the container occur in (3)! (A thicker-walled container can sustain a higher pressure but has a greater weight.) Despite leafing through several tables of physical properties, we could find no material that would fulfil condition (3). None of the materials known today are strong enough, relative to their densities, to be lifted by the upthrust available from releasing a gas stored inside them.

S116 Gravitational acceleration at the surface of the spherical Earth, of radius R, mass M and density ρ, can be written as

$$g = G\frac{M}{R^2} = G\frac{4\pi R^3 \rho}{3R^2} = \frac{4\pi}{3}GR\rho.$$

In order to solve the given problem, we have to find the magnitude of the gravitational acceleration on the surface of a very large disc of thickness H and density ρ, at a point far from the edge of the disc. The result is relatively easy to obtain using an analogy between the two sets of laws governing electrostatic and gravitational interactions.

We draw analogies between a mass m (the 'gravitational charge') and an electric charge q, the gravitational constant G and $1/(4\pi\varepsilon_0)$, and the gravitational acceleration $g = F/m$ and the electric field strength $E = F/q$. In both cases, F is the force experienced by the 'test charge'. We next determine the electric field strength (outside the disc) of an infinitely large disc carrying a homogeneous charge distribution, and then substitute the analogous quantities to obtain the gravitational acceleration for a mass distribution of similar geometry.

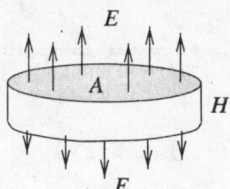

The electrostatic field strength can be calculated by applying Gauss's theorem to a disc of area A,

$$\Phi_E = \frac{1}{\varepsilon_0}\sum q,$$

where $\Phi_E = 2AE$ is the electric flux (*see figure*). If the electric charge density is ρ_q then the total charge surrounded by the closed surface is $\sum q = \rho_q AH$. Therefore

$$2EA = \frac{1}{\varepsilon_0}\rho_q AH,$$

which yields

$$E = \frac{1}{\varepsilon_0} \frac{\rho_q H}{2}$$

for the electric field strength. Substituting the relevant analogues, the gravitational acceleration is found to be $g = 2\pi G \rho H$. This has to be equal to the gravitational acceleration measured at the surface of the Earth, i.e.

$$G\frac{4\pi R\rho}{3} = g_{\text{sphere}} = g_{\text{disc}} = 2\pi G\rho H,$$

which we can write as

$$H = \frac{2}{3}R = \frac{2}{3} \, 6370 \text{ km} = 4250 \text{ km}$$

for the thickness of the 'flat Earth'.

S117 It follows from the similarity of triangles FHC and GKH in Fig. S117.1, that for the short length of rod KH determined by the small angle $\Delta\alpha$, the following equalities hold:

$$\Delta x = GH \, \frac{r}{h} = r\Delta\alpha \, \frac{r}{h} = \frac{r^2}{h}\Delta\alpha.$$

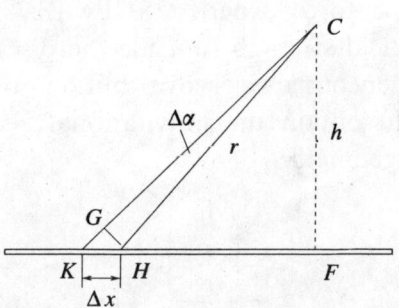

Fig. S117.1

There is a charge $\Delta Q = \sigma\Delta x$ on KH, where σ is the charge per unit length, and the magnitude of the electric field strength corresponding to this charge is

$$\Delta E = \frac{1}{4\pi\varepsilon_0} \frac{\Delta Q}{r^2} = \frac{1}{4\pi\varepsilon_0} \frac{\sigma}{h}\Delta\alpha.$$

This quantity is *independent* of the value of angle α itself. It only depends on the angle $\Delta\alpha$ which the piece of rod subtends at C. Thus the electric field vectors due to the lengths of charged rod situated symmetrically with respect to the bisector of angle C are of the same magnitude and their resultant points in the direction of the bisector (*see* Fig. S117.2).

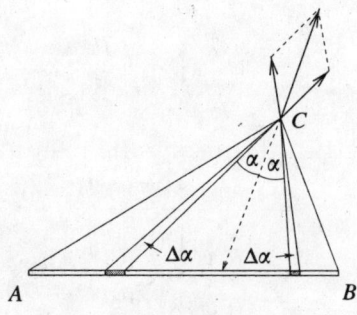

Fig. S117.2

For any given point C there is some particular value of α such that $A\hat{C}B = 2\alpha$, and by superimposing the results for matched pairs of elementary lengths of the rod, the stated result is established.

S118 Using the result of the previous problem, it can be stated that the direction of the electric field at a point on the plane, and a distance h from the end of the infinitely long rod, makes an angle of 45° with the rod.

The magnitude E of the electric field strength can be found using the following 'trick'. Imagine two very long, uniformly charged rods joined end to end. The resultant field strength will be the vector sum of the field strengths of the two 'half-rods' (*see* Fig. S118.1).

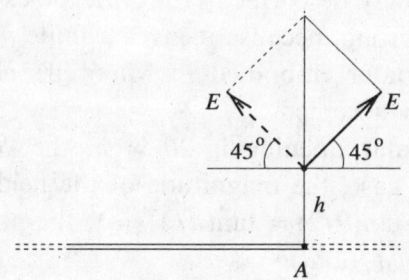

Fig. S118.1

The direction of the resultant will obviously be perpendicular to the rod – in view of the symmetry – and its magnitude, $\sqrt{2}$ times the field strength E for an individual rod, can be found using Gauss's law for electric field lines. Enclose a section of length ℓ of the infinitely long rod in a notional cylinder of radius h (*see* Fig. S118.2).

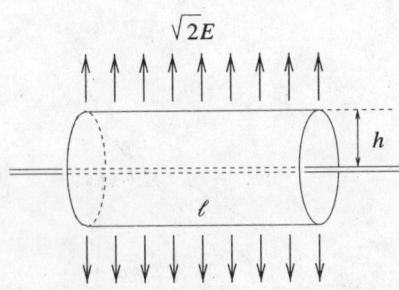

$$\sqrt{2}E$$

Fig. S118.2

There is a charge $Q = \ell\sigma$ inside the cylinder, where σ is the charge per unit length, and the number of field lines crossing the cylinder (the electric flux) is

$$\Phi_E = \sqrt{2}E \, 2\pi h\ell.$$

According to Gauss's law,

$$\varepsilon_0 \Phi_E = Q, \quad \text{giving } E = \frac{\sqrt{2}}{4\pi\varepsilon_0} \frac{\sigma}{h}.$$

S119 (i) When the angular opening approaches 2π, point P is in between the wires and arbitrarily close to them, and so both halves of the current-carrying wire produce very large magnetic fields at P, and in the same direction. Thus in this case the magnitude of the net magnetic field at P approaches infinity. As $\tan(\theta/2)$ also approaches infinity when θ approaches π, Ampère's formula may be correct. However, the expression given by Biot and Savart must be wrong, because it gives a finite value for $B(P)$. In fact, Ampère's result was later embodied in Maxwell's electromagnetic theory, and is now universally accepted.

(ii) When the angular opening is $2\theta = \pi$, the 'V' becomes a straight infinite wire. For this case, the magnitude of the field $B(P)$ is known to be $B = \mu_0 I / (2\pi d)$. Since $\tan(\theta/2) = \tan(\pi/4) = 1$, the proportionality factor in Ampère's formula is $\mu_0 I / (2\pi d)$.

Biot and Savart chose their formula in such a way that it agreed with the expression for the magnetic field due to a straight infinite current-carrying wire already generally accepted. Thus they had as their proportionality factor $\mu_0 I / (\pi^2 d)$.

Note. In the region $\theta < \pi/2$ the difference between the two predictions is relatively small. The ratio of the predicted values for B, $2\theta/\pi \tan(\theta/2)$, shows the greatest difference from unity when $\theta \to 0$ and has a maximum value of $4/\pi$.

S120 (i) If the magnetic field strength at point P is denoted by B_1 then a symmetrically placed second coil, as shown in Fig. S120.1, would also produce a field of strength B_1.

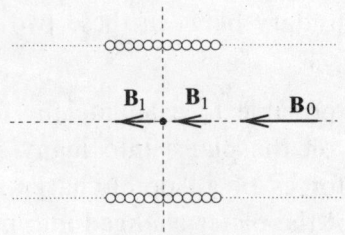

Fig. S120.1

Since the resultant magnitude of the superimposed fields is clearly B_0, it follows that $B_1 = B_0/2$.

(ii) Similar reasoning shows that the horizontal component of the magnetic field vector through P is $B_0/2$; this would be true for any P whose distance from the axis is less than R (Fig. S120.2).

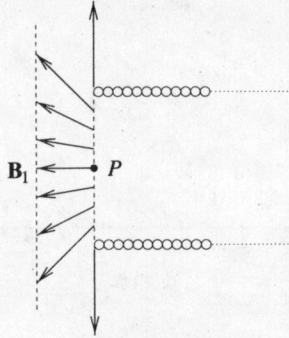

Fig. S120.2

Therefore a total magnetic flux of $\pi R^2 (B_0/2)$ crosses the end of the solenoid. This is exactly half of the flux inside the solenoid; what happens to the other half?

(iii) A qualitative sketch of the field lines can be seen in Fig. S120.3.

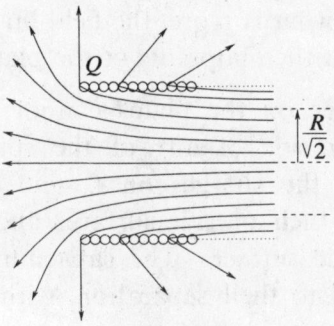

Fig. S120.3

The field line crossing the endmost turn of the solenoid (point Q) continues travelling perpendicularly to the solenoid. Half of the flux travels to the left of this, the rest leaves the coil between the turns. Deep inside the solenoid, the field lines on the boundary between these two halves are at a distance $R/\sqrt{2}$ from its axis.

S121 In theory it is possible to calculate the force between the plates by dividing the charges on the plates into many small point charges and summing the Coulomb forces of all point-charge pairs. If we imagine the positive charges on one of the plates changed into negative ones of the same magnitude, then the magnitudes of the Coulomb forces remain the same, but their signs are reversed. Instead of repulsion, we get attraction between the plates, as for parallel plate capacitors. The energy stored in such a capacitor with plate separation x is $E = Q^2/2C$, where the capacitance C is given by $C = \varepsilon_0 A/x$. Hence $E = Q^2 x/2\varepsilon_0 A$, and the force F is the rate at which this changes as x is changed, i.e. $F = dE/dx$. Hence the force required to hold together plates of area A is

$$F = \frac{Q^2}{2\varepsilon_0 A}.$$

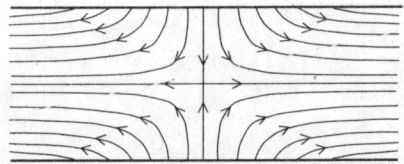

The electric field line structure of the positive–positive plates is very different from that of the positive–negative arrangement. The lines are sketched in the figure. It will be obvious that the electric field is not homogeneous between the plates and, what is more, the field lines do not leave the plates perpendicularly except at the midpoints of the plates.

S122 The net charges on the plates cannot change, but the charges on the plates on either side of any of the spaces must be equal and opposite. Consequently, the charges on C and D must be -1 nC and $+1$ nC, respectively, on their outside surfaces and $+2$ nC and -2 nC, respectively, on their inside surfaces. The capacitance of any pair of plates is inversely proportional to their separation, with 5 mm corresponding to 20 pF.

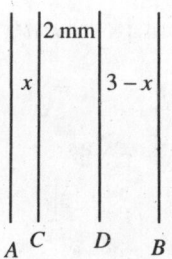

Thus if AC is x mm and DB is $(3-x)$ mm, the capacitances of the three successive capacitors are $100/x$, 50 and $100/(3-x)$ pF. The voltage V_{CD} is therefore 40 V, and V_{AB} is $10x + 40 + 10(3-x) = 70$ V, independent of the value of x.

S123   The charge between the capacitor plates could be notionally divided into two parts, which are then moved away from each other in a direction parallel to the plates. The (induced) charges on the capacitor plates would also move but their totals on each plate would remain unaltered (using the principle of superposition). Continuing in the same manner, the charge Q could be further subdivided until it was 'spread' uniformly on a plane of the same size as, and parallel to, the capacitor plates.

We can now consider two plane capacitors connected in parallel, the distances between their plates being x and $d-x$. Since one plate of each of the capacitors is earthed, and the other is common, the voltages between the pairs of plates are identical. Their electric field strengths are therefore inversely proportional to their plate separations, i.e. $E_1/E_2 = (d-x)/x$. The electric flux emanating from the charge Q is divided in the same proportion (Gauss's law), and the ratio of the charges on the earthed plates is the same as well. Since a total charge of $-Q$ accumulates on the plates, respective charges of

$$Q_1 = -Q\frac{d-x}{d} \quad \text{and} \quad Q_2 = -Q\frac{x}{d}$$

accumulate on the two separate plates.

S124   Very far from the capacitor the electric field is determined by the total electric charge of the system. The electric field outside the capacitor is zero; therefore the total electric charge (the sum of the charges on the plates) must also be zero. That means $Q_1 = -Q_2 = Q$.

The long distance behaviour of the electric field of an electrically neutral system is determined by its total electric dipole. In our case, taking the component of this dipole moment normal to the plates gives

$$p - Qd = 0,$$

which gives the charges on the plates as

$$Q = \frac{|\mathbf{p}|}{d}.$$

Thus the charges on the plates do not depend upon the position of the dipole.

Note. (i) The same method can also be applied in P123. The relevant equations are:

$$Q_1 + Q_2 + Q = 0,$$
$$Q_1 x - Q_2 (d - x) = 0,$$

and their solution is

$$Q_1 = -Q(d-x)/d, \quad Q_2 = -Q x/d.$$

(ii) Alternatively, we can use the result from the case of a single charge by placing two charges of opposite signs into the parallel plate capacitor and applying the superposition principle. This is a simple way to find the solution of the general problem, in which the dipole momentum is not necessarily perpendicular to the plates.

S125   Figure S125.1 shows a light ray passing through succesive plano-parallel plates of different refractive indices.

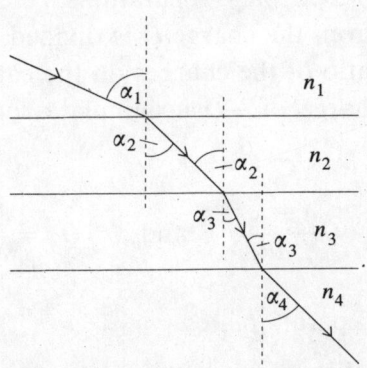

Fig. S125.1

According to Snell's law,

$$\frac{\sin \alpha_2}{\sin \alpha_1} = \frac{n_1}{n_2}, \quad \frac{\sin \alpha_3}{\sin \alpha_2} = \frac{n_2}{n_3}, \quad \ldots$$

It can be seen that the product of the sine of the angle of incidence and the absolute refractive index has the same value at all interfaces, i.e. $n \sin \alpha$ is constant along the light ray's trajectory.

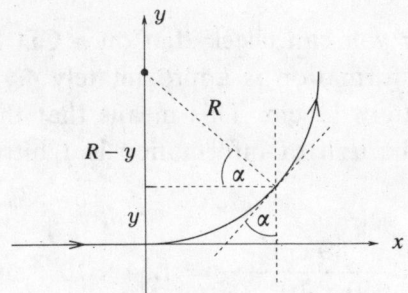

Fig. S125.2

This relationship is also valid for a medium whose refractive index continuously changes in one direction, since the medium can be considered as consisting of thin plano-parallel plates. Place the origin of the coordinate system at the point where the light ray enters the medium. In this case, the angle of incidence for the first 'plate' starting at $y = 0$ is 90° and the refractive index is n_0, which gives the above constant as $n(y) \sin \alpha = n_0$.

The light travels along a circular arc of radius R and we first examine its relationship to coordinate y. From Fig. S125.2 it is clear that

$$n_0 = n \sin \alpha = n(y) \frac{R - y}{R}.$$

This gives the space-dependence of the refractive index as

$$n(y) = \frac{R}{R - y} n_0.$$

The material with the greatest known refractive index is diamond, but even the refractive index of this material does not reach the value $n_{max}' = 2.5$. It is this limit that sets the maximum angular size of the arc the light ray can cover. If the refractive index changes from $n_0 = 1$ to $n_{max} = 2.5$ then the maximum value of y is $\frac{3}{5}R$, corresponding to an arc of angular size 66.4°.

In practice, it is difficult to constrain light to a circular arc. However, it is possible to make up solutions in which the concentration of solute, and therefore the refractive index, shows a continuous vertical change. For such a medium, the light ray does not propagate along a circular arc but follows some other continuous curve.

Note. Why does the light ray entering along the x-axis start to bend at all? The reason is that there is no such thing as an infinitely thin light ray; a 'ray' always implies a beam of finite width, with the refractive index and, hence, the speed of propagation varying across the beam profile. As a consequence, the wavefront becomes non-planar and the beam bends.

S126 Using a ruler you can check that on a CD the inner diameter of the area for storing information is approximately 4.4 cm, whilst the outer diameter is approximately 11 cm. This means that the useful surface area is about 80 cm^2. As the unit of information is 1 bit (1 byte = 8 bits) the surface area per bit is

$$A = \frac{80 \text{ cm}^2}{650 \times 10^6 \times 8} = 1.54 \times 10^{-8} \text{ cm}^2.$$

Assuming that the 'shape' of a single unit of information is a square ($A = a^2$), the linear size of a bit is $a = 1.24 \times 10^{-4}$ cm = 1.24 μm.

The information on a CD is stored in a very long spiral starting near the centre and working outwards; this is the reverse of what happens on traditional LPs. A small piece of a CD can be used as a reflection grating and you can measure its diffraction pattern using a laser beam of known wavelength. The effective grating spacing is the width of the grooves and it is simplest to use normal incidence and measure the distance between the two first-order interference maxima on a screen. The typical wavelength of an optical laser is 670 nm and it is convenient to choose 1 m as the distance between the CD and the screen.

The condition for the first-order intensity maximum in the interference pattern of a grating is $d \sin\theta_1 = \lambda$, where d is the grating spacing. Experimentally, $\theta_1 \approx 25°$, and inserting the data into the grating equation gives the grating spacing as $d \approx 1570$ nm = 1.57 μm. Although close, this estimate is somewhat larger than our previous one. The difference between the two results is not a measurement error, but a consequence of the separating walls between neighbouring grooves. The effective width of the information is only about 0.5 μm. This means that the total effective area of information is only some (80/3) cm$^2 \approx 27$ cm^2, and that 1 bit has a rectangular shape approximately 1 μm long.

Note. The information density on CDs is uniform, but the rotation rate changes according to the position of the reading head.

S127 $n\lambda = d\sin\theta = (10^{-3} \text{ m}/300)\sin 24.46° = 1380$ nm and the only possible values for n and λ to put the red and blue/violet light into the appropriate parts of the spectrum are $n_{\text{R}} = 2$, $\lambda_{\text{R}} = 690$ nm and $n_{\text{BV}} = 3$,

$\lambda_{\text{BV}} = 460$ nm. In all physically possible cases

$$n\lambda \leq d \sin 90° = 3333 \text{ nm},$$

and the only other pair of integers which are in the ratio $3m : 2m$, with m less than $(3333/1380) = 2.4$, is 6 and 4. Thus there is only one more angle at which a two-component line will be observed; i.e. at

$$\sin^{-1} \frac{(6 \times 460)}{3333} = 55.9°.$$

S128 (i) When the monochromatic laser beam falls on the diffraction grating at normal incidence, the positions θ of maxima of the interference pattern are given by

$$d \sin \theta = m\lambda, \qquad (m = 0, \pm 1, \pm 2, \pm 3, \ldots),$$

where d is the grating spacing and λ is the wavelength of the laser beam. If the grating is rotated through an angle ϕ around an axis parallel to the lines of the grating, we have to modify the above equation. It is enough to consider the interference from only two slits of the grating, as shown in Fig. S128.1.

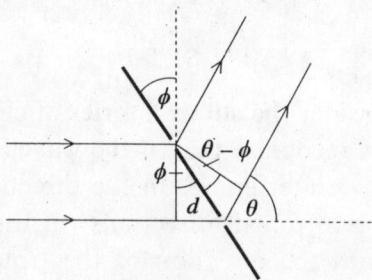

Fig. S128.1

The optical path difference consists of two parts:

$$\Delta_1 + \Delta_2 = d \sin \phi + d \sin(\theta - \phi).$$

Thus the modified equation for the interference pattern is

$$d[\sin \phi + \sin(\theta - \phi)] = m\lambda.$$

The principal result is that the pattern becomes asymmetrical, with only the position of the zeroth-order maximum remaining unchanged. If ϕ is counted as positive when the grating is rotated anticlockwise (and θ is counted as positive in the same sense), then the density of interference maxima becomes larger at positive angles, whilst it decreases for negative ones. If the lines of the grating are vertical, all the interference maxima lie along a horizontal straight line; this is to be contrasted with what happens in (ii).

Note. The naïve idea that the effect of rotating the grating is to decrease the 'effective' size of the grating spacing is *false*.

(ii) Since the diffraction grating is an array of a large number of identical parallel slits, as a first step it is enough to investigate the diffraction from a single slit. If the beam falls on a very narrow slit at normal incidence, the diffraction pattern is a weak straight line. If the slit is a little wider, interference minima along this line can be identified.

When the slit is tilted 'forward' (i.e. rotated around a horizontal axis which is perpendicular to the slit as well as to the direct beam), the zeroth-order maximum remains unchanged. Though the incoming beam strikes different parts of the slit with different phases, beyond the slit, in the direction of the direct beam, there are no phase differences (Fig. S128.2) and no change in the pattern results.

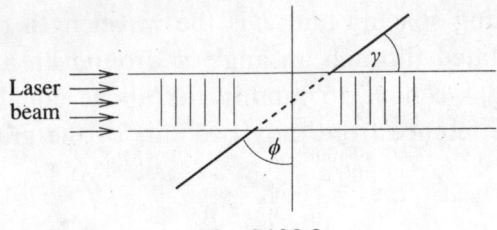

Fig. S128.2

It may be easier to consider the slit as a series of closely spaced very small holes. Now, we need only recognise that, if the wavelets originating from the holes produce constructive interference in the direction of the direct beam, then the same must be true for all directions (in three dimensions), which make the same angle with the direction of the (rotated) slit. If the slit is tilted by an angle ϕ, then the angle between the slit and the direct beam is $\gamma = 90° - \phi$. The same angle γ between the slit and a diffracted ray occurs for any ray that lies on the cone which has semi-angle γ and the direction of the slit as its axis. On the screen we would see a plane section of this cone (*see* Fig. S128.3).

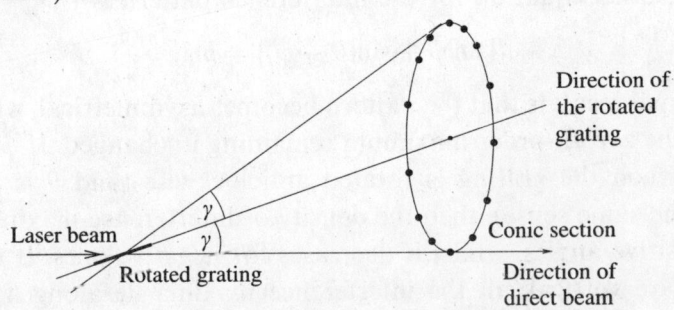

Fig. S128.3

The apex of the cone is the midpoint of the slit, and its semi-angle is γ. The conic section can be ellipse, parabola or hyperbola. A parabola is obtained if $\gamma = \phi = 45°$.

Returning to the original problem, we conclude that the interference pattern of the tilted grating consists of bright spots lying on a conic section.

S129 Because of the surface tension of the liquid, its height between the two objects is not the same as it is outside the objects; in the case of (*a*) water it is higher, whilst for (*b*) mercury (which has a negative angle of contact) it is lower.

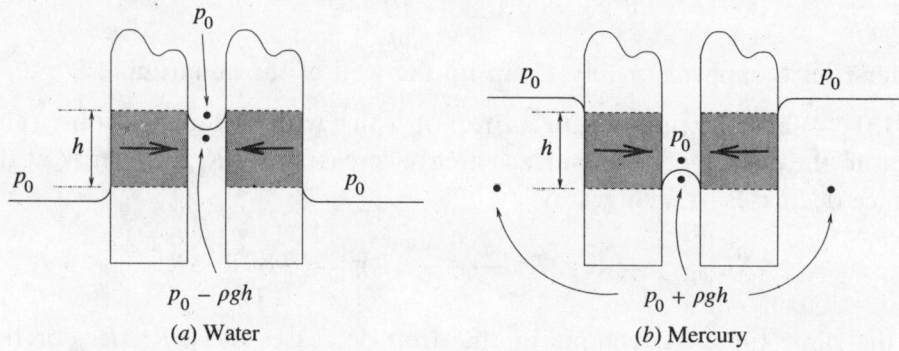

(*a*) Water   (*b*) Mercury

Just above the liquid surface between the objects the pressure has to be atmospheric in both cases and, correspondingly, just below the surface it has to be less than atmospheric for water and more than atmospheric for mercury. As can be seen from the figure this leads to net inward forces (acting on the shaded areas) in both cases and a tendency for the objects to move towards each other.

S130 The pressure of the water changes linearly with the increase in height. At the bottom of the meniscus it is equal to the external atmospheric pressure p_0, and at the top to $p_0 - \rho g h$. The average pressure exerted on the wall is $p_{\text{average}} = p_0 - \rho g h / 2$. The force corresponding to this value, for an aquarium with side walls of length ℓ, is $F_1 = \ell p_{\text{average}} h$.

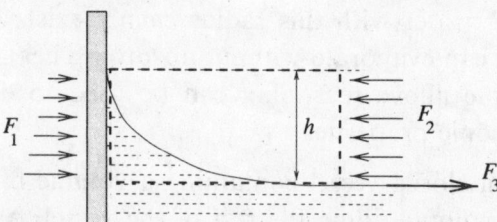

Consider the horizontal forces acting on the volume of water enclosed by the dashed lines in the figure. The wall pushes it to the right with force F_1, the external air pushes it to the left with force $F_2 = \ell p_0 h$, and the surface tension of the rest of the water pulls it to the right with a force $F_3 = \ell \gamma$. The resultant of these forces has to be zero, since the volume itself is at rest. This means that

$$\left(p_0 - \frac{1}{2}\rho g h\right)\ell h - p_0 \ell h + \ell \gamma = 0,$$

which we can write as

$$h = \sqrt{\frac{2\gamma}{\rho g}} = \sqrt{\frac{2 \times 0.073}{1000 \times 10}} = 0.0038 \text{ m}.$$

Water rises by approximately 4 mm up the wall of the aquarium.

S131 When the radius R of a drop of water with surface tension γ (also equal to the energy per unit surface area) decreases by ΔR, the energy of the surface decreases. It changes by

$$\Delta E_{\text{surface}} = 4\pi\gamma \left[R^2 - (R - \Delta R)^2\right] \approx -8\pi R\gamma\Delta R.$$

At the same time, the volume of the drop decreases by $4\pi R^2 \Delta R$. For this quantity of water to evaporate, energy

$$\Delta E_{\text{evaporation}} = 4\pi L\rho R^2 \Delta R$$

has to be supplied. Here ρ is the density of water and L its latent heat of evaporation.

The decrease in the surface energy could provide the evaporation energy of the drop if $|\Delta E_{\text{surface}}| > \Delta E_{\text{evaporation}}$, i.e. if

$$R < \frac{2\gamma}{\rho L} \approx 7 \times 10^{-11} \text{ m}.$$

Since this radius is of the same order of magnitude as the size of one water molecule, a drop of water with this radius *cannot* exist. Therefore, there is no water drop that can evaporate without absorbing heat, or losing internal energy. However, the above reasoning can be used to estimate molecular sizes using macroscopic properties.

S132 Consider a closed vessel containing a volume of liquid, with saturated vapour of this liquid filling the rest of the vessel. As illustrated in the figure, let a capillary tube of radius r be immersed in the liquid, which does not wet its walls.

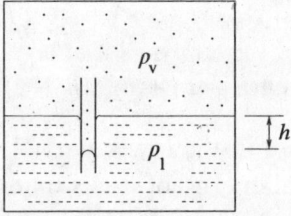

In such circumstances, the level in the capillary tube falls to a depth h below the liquid surface. The magnitude of h can be found using the equilibrium relationship between hydrostatic force and the surface tension of the liquid $2\pi r\gamma = \rho_1 g h\,\pi r^2$, where γ is the surface tension of the liquid and ρ_1 is its density. This gives $h = 2\gamma/(\rho_1 g r)$.

The liquid is in equilibrium with its saturated vapour, both in the capillary tube and at the plane surface of the liquid. In the capillary tube, however, the pressure of the vapour is a little higher at the interface. The difference is caused by the pressure of the vapour column, of height h and density ρ_v, above it. It follows that

$$\Delta p_v = \rho_v g h = \frac{2\gamma}{r}\frac{\rho_v}{\rho_1}.$$

As $\rho_v \ll \rho_1$, this difference between the pressures is much smaller than the pressure of curvature corresponding to the radius r, but, given long enough, it is sufficient to bring about the phenomenon described.

S133 Let A denote the cross-sectional area of the piston and y the vertical displacement between its initial and final equilibrium positions (*see figure*).

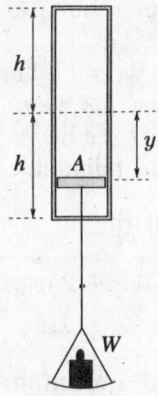

The decrease in potential energy of the weight W increases the internal energy of the air inside the cylinder. Conservation of energy between the initial and the final states gives

$$Wy = \frac{5}{2}\left[p_1 A(h - y) + p_2 A(h + y) - 2p_0 Ah\right], \tag{1a}$$

where p_1 is the final pressure in the lower part of the cylinder and p_2 that in the upper part. The internal energy of a gas made up of diatomic molecules has been written in the form $\frac{5}{2}pV$. If W is very large, the decrease in its potential energy (and the corresponding increase in the internal energies of the gases) is very large, and the initial internal energy of the air can be neglected. Thus

$$Wy = \frac{5}{2}\left[p_1 A(h - y) + p_2 A(h + y)\right]. \tag{1b}$$

When the load is finally at rest,

$$(p_1 - p_2)A = W. \tag{2}$$

The temperatures and the masses of the gases in the two halves are identical, and so their internal energies must be equal:

$$\frac{5}{2}p_1 A(h - y) = \frac{5}{2}p_2 A(h + y). \tag{3}$$

Equations (1b), (2) and (3) yield $y = \sqrt{5/7}h$ for the displacement of the piston, i.e. the gas in the lower part is compressed to $1 - \sqrt{5/7} \approx 15$ per cent of its original volume.

> *Note.* The surprising result is that the volume of air in the lower part does not tend to zero, however large the weight is, even though gases are supposed to be compressible! The large load increases the internal energy, and hence the temperature, of the enclosed gas. This causes considerable increases in not only the absolute pressures, but also in the difference between the upper and lower pressures.
>
> In practice, the applied weight is limited by the mechanical load-bearing capacity of the structure and the melting points of the materials used. We should also consider whether the increased temperature is one at which air can still be treated as an ideal diatomic gas.

If the initial internal energy of the air is *not* neglected then the result is

$$\frac{y}{h} = \frac{\sqrt{35W^2 + 25p_0^2 A^2} - 5p_0 A}{7W}.$$

S134 Consider a 'box-shaped' mountain of average density ρ, base area A and height h. In order to melt its bottom layer of thickness d and specific latent heat L, energy $Ad\rho L$ would be required. The total mass of the mountain is approximately $Ah\rho$, and the energy released if it sank a distance

d would therefore be $Ah\rho gd$. The base of the mountain does not melt under its load if

$$Ad\rho L > Ah\rho gd, \qquad \text{i.e. } h < \frac{L}{g}.$$

Approximating the required latent heat by the latent heat of melting of metals ($200 - 300$ kJ kg^{-1}), we estimate the maximum possible height of mountains on Earth to be 20–30 km. This is of the right order of magnitude. Allowing for the fact that the base of the mountain does not actually have to melt, but rather that the size of mountains is limited by the yield strength of their constituent materials, the estimated height of the highest mountains on Earth is surprisingly accurate.

Gravitational acceleration is significantly smaller on Mars than on Earth ($g_{\text{Mars}} \approx 4$ m s^{-2}). Therefore mountains, consisting of similar rocks, could be higher on Mars than on Earth. Indeed, the highest mountain on Mars, Mons Olympus, is 26 km high!

S135 In order to simplify the calculation, choose a system of units in which the initial volume and the external pressure are unity, and the units of the number of moles and the gas constant have a product that is also unity. This reduces the usual ideal gas equation $pV = nRT$ to $pV = T$. In this system, the molar heat has to be multiplied by the gas constant to yield the normal molar heat capacity.

The initial pressure of the air is given by the sum of the pressure of the 76 cm-high column of mercury and the atmospheric pressure, a total of 2 units. Any increase in length of the air column implies a corresponding decrease in the mercury one. Therefore as the air expands from 1 to 2 units, its pressure decreases linearly from 2 to 1 units as shown in Fig. S135.1.

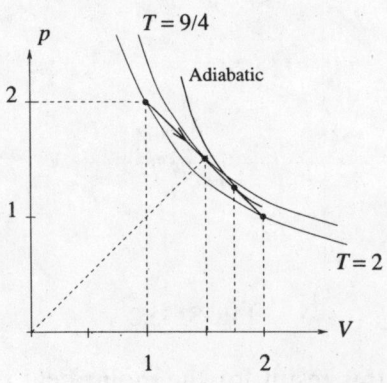

Fig. S135.1

The temperature of the enclosed air is initially 2 units, and is still 2 units at the end of the straight section shown in the figure, since both points lie on the isothermal $pV = 2$. In our representation, the isothermals are symmetrical about the bisector of the axes (i.e. about the line $p = V$). Thus the highest temperature reached corresponds to the isothermal to which the straight-line segment is tangential. As shown in Fig. S135.1, this occurs at the middle of the process, when $p = \frac{3}{2}$ and $V = \frac{3}{2}$. The maximum temperature is therefore $\frac{9}{4}$.

The equation of the straight line representing the process is $p = 3 - V$. Applying the first law of thermodynamics to a typical section of the line gives

$$\frac{5}{2}\Delta T = C\,\Delta T - p\,\Delta V,$$

where the left-hand side of the equation is the change in the internal energy of the air (diatomic molecules have, $C_V = \frac{5}{2}R$), and on the right-hand side, C is the molar heat in question. Expressing p as $p = 3 - V$ in the equation $pV = T$, gives $(3 - V)V = T$. From this it follows that small changes in V and T are connected by $(3 - 2V)\Delta V = \Delta T$. The change in the internal energy of the air then becomes

$$\frac{5}{2}\Delta T \doteq C\,\Delta T - \frac{3 - V}{3 - 2V}\,\Delta T,$$

which simplifies to

$$C = \frac{21 - 12V}{6 - 4V}.$$

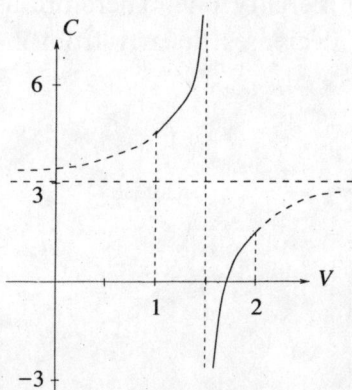

Fig. S135.2

Figure S135.2 shows this result for the molar heat plotted against volume. The curve is singular (the molar heat approaches infinity) at $V = 3/2$ because at this point the straight line is a good approximation to the isothermal corresponding to the maximum temperature. During an isothermal process, heat is transferred but the temperature does not change, and this is why the molar heat tends to infinity. Beyond this point, the molar heat becomes negative, which means that in spite of the positive heat transfer, the internal energy of the air decreases because the work done by it in expanding is greater than the heat transferred.

The most interesting part of the process occurs when $V = \frac{7}{4}$, i.e. when only a quarter of the mercury is left in the tube. At this point the molar heat is zero and implies that the process is adiabatic, since there is a change of temperature without any heat transfer; at this point, the straight line describing the process is tangential to an adiabatic curve. Beyond this point, the temperature decreases further but the molar heat becomes positive; this can only correspond to the system being cooled.

If more heat is given to the system, the representative point no longer follows the straight line but continues along the adiabatic one. The temperature of the air does decrease, but not by as much as the straight line would suggest. The decrease in the internal energy of the air is equal to the work done against atmospheric pressure (whilst lifting the mercury) and in accelerating the ejected mercury.

Returning to the original question, that of the heat transfer necessary to push the mercury out of the tube, it can be seen that it is only necessary to transfer heat whilst V is in the range $V = 1$ to $V = \frac{7}{4}$. Simple arithmetic shows that the work done by the air up to this point is (in this system of units) $\frac{39}{32}$, whilst the increase in internal energy is $\frac{15}{32}$. The required heat transfer is therefore $Q = \frac{27}{16}$. In standard units, this means that if the column of mercury originally enclosed n moles of air, then the heat transfer required to remove it is $\frac{27}{16} nR$.

S136 The molten magma came into contact with the ice, and huge volumes were melted at the base of the 500-m-thick ice cap. As the density of water is greater than that of ice (i.e. the volume of meltwater is smaller than that of the ice from which it was formed), and moreover some part of the water could flow away, a huge conical hole would have formed under the ice. But the extremely heavy ice above the hole sank, leaving the depression at the surface. Under the ice crater we could have found the recently solidified magma intrusion, the meltwater in a conical cavity and the ice cover (*see figure*). The amount of meltwater depends on the quantity of magma, but the shape of the ice crater is determined by the hydrostatic pressure of the ice and water.

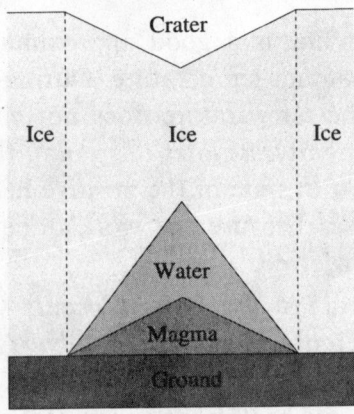

The eruption broke through the ice cap on the second day and a hot, black ash-cloud, 500 m high, was formed. This was carried up to an altitude of 3000 m by the buoyancy force of the cold air. At the end of two weeks the cloud column had become white and reached a height of 10 km.

Altogether the eruption melted 3 km^3 of ice in two weeks. The meltwater flowed under the ice of the glacier into a lake situated within a nearby volcanic depression, the Grimsvotn caldera. A deep depression, 8 km long and a few hundred metres wide was formed in the ice surface at the eruption site. The rate of melting was extremely high, 0.5 km^3 per day for the first four days. At the same time a new mountain, 0.6 km long and 150–300 metres high, was built under the ice by the eruption.

The meltwater was held in Grimsvotn caldera under the glacier for five weeks, before it escaped. A gigantic wave swept across part of the south-east lowlands, destroying everything (roads, bridges, etc.) in its path.

S137   Assume that the cavity is located at a depth h below the surface. When the flue is filled with water, the pressure in the cavity is $p = p_0 + \rho g h$, where p_0 is the atmospheric pressure and ρ is the density of water. Boiling starts at that depth when the temperature is such that the pressure of saturated water vapour ($Ae^{-L_m/(RT)}$) is equal to p. The value of the molar heat of vaporisation L_m occurring in the formula is approximately 40 kJ mol^{-1} and that of the molar gas constant R is 8.3 J mol^{-1} K^{-1}. Using these data, the constant A can be determined from the boiling point of water (373 K) at atmospheric pressure; the result is $A = 4.1 \times 10^{10}$ Pa.

For the purposes of the calculations, the temperature of the ground at the surface can be assumed to be $T_0 = 290$ K and, as it increases by one degree per metre, the temperature at depth h is given by $T = T_0 + h$. The following equation can now be formulated, using the equality of the pressures when the water starts to boil:

$$p_0 + \rho g h = A e^{-\frac{Lm}{R(T_0+h)}}.$$

This transcendental equation can be solved by numerical methods, i.e. by carefully considered guesses! The solution is $h = 197(\approx 200)$ m. The pressure at such a depth is approximately 20 atm and the temperature 487 K = 214 °C.

When boiling starts, the first steam bubbles rise and push the water out of the flue, the pressure decreases to 1 atm and the superheated water at 214 °C starts to boil violently, producing as much steam as is necessary to reduce its temperature to 100 °C. The temperature drop of $\Delta T = 114$ °C implies an excess internal energy of $cm\Delta T$, where c is the specific heat of water and m is the mass of water in the cavity. This surplus energy is sufficient to produce $m_s = 44$ tons of steam, implying

$$cm\Delta T = Lm_s,$$

where L is the heat of vaporisation of water. After substituting the relevant data, we find the mass of the water in the cavity to be $m = 207$ tons. Since the density of water is only around 850 kg m^{-3} at such a high temperature, this gives an approximate value of 244 m^3 for the volume of the cavity.

S138 The heat conducted away in a given time must equal (minus) the latent heat of fusion of the additional ice formed in that time. Thus for an area A of the lake

$$\lambda_i A \frac{\Delta T}{x} = L_i \frac{\mathrm{d}}{\mathrm{d}t}(\rho_i A x).$$

Simple integration gives that $t = \frac{1}{2}x^2 B$, where $B = (\rho_i L_i)/(\lambda_i \Delta T)$. Inserting the relevant numerical values gives the time as about 90 hours.

S139 According to the law of heat conduction the heat transferred is directly proportional to the thermal gradient (temperature difference divided by distance), the area and the time. So

$$\text{Time} \propto \frac{\text{Heat capacity}}{\text{Area} \times \text{Temperature gradient}} \propto \frac{L^3}{L^2 L^{-1}} = L^2.$$

Since $M \propto L^3$, the time is proportional to $M^{2/3}$ and a mammoth should take $2(8000/5)^{2/3}$ days, i.e. about nine months, assuming that the turkey and the mammoth both start to thaw from the same temperature and are defrosting in similar environments. (In fact, the Siberian summer is too short to defrost a mammoth.)

S140 Consider first the state of the ice and the pressure in the container at 100 °C. Since the density of saturated water vapour is 0.5977(≈ 0.6) kg m^{-3}

and its pressure is 1 atm, the 0.6 kg of ice is completely transformed into vapour at 100 °C, indeed into saturated vapour at a pressure of 1 atm.

How does ice, at a temperature of −10 °C, turn into saturated water vapour at 100 °C? If the temperature is increased very slowly, the system passes through a number of equilibrium states. First, the ice sublimes and the ice phase is in equilibrium with the vapour phase. This lasts until the temperature and pressure of the triple point (0.01 °C, 610 Pa) is reached. At the triple point, a liquid state appears alongside the ice and water vapour. Further heating makes the solid phase disappear, and only water and saturated water vapour remain in the container. It is interesting to note that subsequently the water boils steadily at 100 °C, until all the water has been transformed into vapour.

From the point of view of heat absorption, only the initial and the final states are important. The heat Q absorbed by the system, as it passes with increasing internal energy through the 'ice–water–vapour' states, is independent of the intermediate states. For calculational purposes, the heating of 0.6 kg of ice should be divided into four stages (warming the ice, melting the ice, warming the water and boiling the water) to give:

$$Q = c_i m \Delta T_1 + L_f m + c_w m \Delta T_2 + L_v m,$$

where $c_i = 2.1$ kJ kg^{-1} °C^{-1} is the specific heat of ice, $\Delta T_1 = 10$ °C, $L_f = 334$ kJ kg^{-1} is the heat of fusion of ice, $c_w = 4.2$ kJ kg^{-1} °C^{-1} is the specific heat of water, $\Delta T_2 = 100$ °C and L_v is the specific heat of vaporisation of water.

The specific heats of ice and water are slightly dependent on temperature and pressure, and the heat of fusion of the ice also depends a little on pressure. But these variations are small and can be neglected. However, the situation is different for the heat of vaporisation of water. It is true that the heat of vaporisation only varies a little with temperature, but its dependence on pressure is significant.

The value usually given in tables, $L_v = 2256$ kJ kg^{-1}, covers, not only the higher internal energy of the vapour, but also the work done against atmospheric pressure. In the present problem, this work was done when the container was evacuated, i.e. the heat to be transferred to the system is smaller by this amount, which is $-p \Delta V = -101.3$ kJ. As this figure is that for 0.6 kg of water,

$$L_v = \left(2256 - \frac{101.3}{0.6}\right) = 2087 \approx 2090 \text{ kJ kg}^{-1}$$

is the more accurate value to be used for the present calculation. After substitution of the data, the heat transfer is found to be $Q = 1720$ kJ.

Neglecting the work done against atmospheric pressure would have introduced an error of nearly 6 per cent, while ignoring the slight dependence of the other coefficients on temperature and pressure causes an inaccuracy of approximately 1 per cent. The main reason for this is that the normal change in the volume in the course of a water–vapour transition is very significant (a factor of about 1600). The volumes of the water and ice are negligible compared with that of the vapour, and they are tacitly neglected in the solution.

S141 Water continues to vapourise in the closed container until the space above it is saturated. The total pressure above the liquid, therefore, is the sum of the pressures of the saturated vapour and the air enclosed in the container, i.e. it is always higher than the pressure of the vapour alone. This means that the water *cannot start boiling at any temperature.*

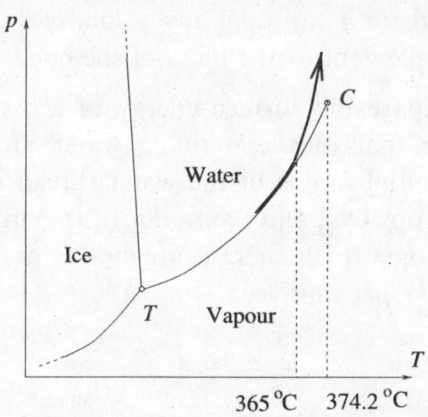

If the container is strong enough, it can withstand a large rise of the temperature. We can ask the question: up to what temperature could liquid water still be found in the container? Data tables for saturated water vapour tell us that the critical temperature of water is 374.2 °C. Its critical density is 326.2 kg m^{-3}, which means it is not possible for all the water to evaporate before the critical point C is reached. If it did, the density of water in the container would have to be around 500 kg m^{-3}. Indeed, a strange change of state occurs at lower temperatures! The density of water is 500 kg m^{-3} at approximately 365 °C, but the density of saturated vapour is 160 kg m^{-3} at this temperature. This means that at 365 °C, liquid water fills the whole container. The explanation for this is that, as the temperature rises in this range, the density of water decreases more rapidly than the density of water vapour increases.

At temperatures above 365 °C, the water in the container remains in liquid state, but its pressure increases. As shown in the figure, the system departs from the boiling curve and continues above it, in the region of the phase diagram corresponding to the liquid phase. The terms 'liquid' and 'vapour' lose their meaning above the critical temperature; rigorous application of the terminology would imply that water is then in a gaseous state.

In what state is the air in the container? In the relevant temperature range (360–380 °C), its pressure is around 200–300 atm. At such a pressure, part of the air is dissolved in the water, and the rest is compressed into very small bubbles filling only 0.1–0.2 per cent of the container. Thus, the air has no noticeable effect on the behaviour of the water.

S142 Only the energy of the air molecules can be relevant and so T appears in the combination kT. Dimensional analysis using ℓ, F, m and kT shows that the amplitude must depend on combinations of variables of the form $(kT/F)^q \ell^{1-q}$, where q can take any value; m does not appear and so the amplitude is independent of the mass of the cobweb.

S143 Let us compare the surface energy of a cylindrical water thread on the cobweb with that of the periodic water drops formed from the thread. Denote the initial radius of the water thread by r, the 'wavelength' (separation) of the drops by λ and the radius of the drops by R, all as shown in the figure. We can ignore all energies (including gravitational) other than the surface energy of γ per unit area.

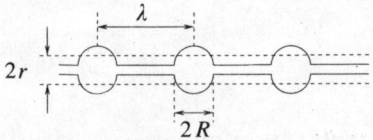

Initially the surface energy of the cylindrical water thread of length L is

$$E_1 = 2\pi r L \gamma.$$

Ultimately L/λ drops, each of radius R, are formed, and their surface energy is given by

$$E_2 = 4\pi R^2 \frac{L}{\lambda} \gamma,$$

where we have ignored the thickness and surface area of the cobweb threads themselves.

The radius R is determined from the conservation of matter.

$$\pi r^2 L = \frac{4\pi R^3}{3} \frac{L}{\lambda}.$$

During drop formation the surface energy must decrease, $E_2 < E_1$. Eliminating R from these equations, we obtain $\lambda > \frac{9}{2}r$. This result shows that the 'wavelength' of the drops must be larger than a certain critical value λ_{crit}, which is proportional to the initial radius of the water thread.

> *Note.* (i) A Belgian physicist, *Joseph A. F. Plateau* (1801–1883) first showed that the critical 'wavelength' is larger than stated above. He found that $\lambda_{\text{crit}} = 2\pi r$ by investigating the constriction of the water thread caused by surface tension. He assumed a periodical change in the diameter of the thread and considered the effect of pressure differences associated with the curvature of the liquid surface.
>
> (ii) The Nobel-laureate English physicist, *John W. S. Rayleigh* (1842–1919) investigated the stability of the water thread and, according to his very careful and detailed calculations on drop formation, the 'winning wavelength' is $\lambda_{\text{win}} = 9.02r$.

S144 (i) Early on, particles coming from the right, and rebounding elastically, transmit a greater momentum to the body than those colliding inelastically from the left. For this reason, a resultant force acting to the left accelerates the body. The faster the body moves to the left, the lower the rate at which particles collide with it from the right, and the lower their relative velocity when they do. For those impinging from the left the converse is true, and the net force acting on the body decreases with time.

After a sufficiently long time, the body moves at uniform speed v_1. The condition for this equilibrium situation is that, in unit time, the particles impart the same momentum from the right as from the left.

The particles from the right reach the body with a relative speed $v_0 - v_1$ and rebound with the same relative speed. In a short time interval Δt, the particles coming from within a distance of $(v_0 - v_1)\Delta t$ reach the cylinder, and each imparts an impulse proportional to $2(v_0 - v_1)$ to the body. Thus the force from the right is proportional to $2(v_0 - v_1)^2$. Similarly, particles inelastically colliding from the left at a relative speed of $(v_1 + v_0)$ produce a force proportional to $(v_0 + v_1)^2$. The condition for constant speed is

$$2(v_0 - v_1)^2 = (v_0 + v_1)^2, \qquad v_1 = \frac{\sqrt{2} - 1}{\sqrt{2} + 1} v_0 \approx 0.17\, v_0.$$

(ii) Assume that the colliding particles are the molecules of a gas at a certain temperature. If the body continued to move uniformly, even after a very long time, it would imply that a *perpetuum mobile* (perpetual motion machine) of the second kind were possible, i.e. a heat engine could be built that could continually extract energy from a single heat reservoir. All experience indicates that such an engine cannot be made.

Where is the mistake in the above reasoning? The heat produced by the inelastic collisions has not yet been taken into account! The particles inelastically bombarding the left end of the cylinder heat it up. If the heat so produced is continuously removed (i.e. the body is cooled), then the motion described in part (i) is sustainable. The result is a normal heat engine working between two heat reservoirs, the gas of bombarding particles and the cooling medium.

However, if the body is not cooled, sooner or later it warms up. The molecules forming the sides of the warm body vibrate at the average speed corresponding to its temperature, and the gas particles rebound from it, sometimes at a greater speed, and sometimes at a lower speed, than they would from a colder body. The result is that finally the body cannot absorb any more heat from the gas at either end (or it would warm up further). Then the collisions are effectively elastic on both ends – taking time averages – and so the impulsive forces are equal. Thus, after a very long time, when thermal equilibrium has been reached, the body has to stop.

S145   As the space probe is very far from the solar system we may neglect the solar and cosmic background radiation. Without any protecting shields, the heat production of the nuclear energy source is radiated away by the surface of the space probe according to the Stefan–Boltzmann law:

$$I = \sigma A T^4,$$

where σ is the Stefan-Boltzmann constant, A is the surface area of the space probe and T is its surface temperature. When a thin protecting shield encloses the space probe, the same radiation process occurs at the outer surface of the shield, and so the temperature of the shield must be T. However, the shield also emits inwards, and consequently the surface of the probe absorbs an amount of radiation equivalent to that radiated into space (*see figure*). This means that the surface of the probe must re-radiate a total received intensity of $2I$ at a new temperature T_1, where

$$2I = \sigma A T_1^4.$$

It follows that $T_1 = \sqrt[4]{2}\, T$.

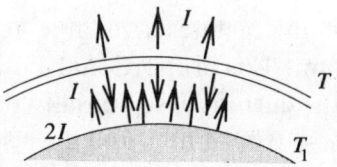

For N protecting shields, the net radiation through them will still be I. Repeated application of our previous argument shows that the space probe radiates $(N + 1)I$ and implies that the temperature of the surface of the probe is $T_N = \sqrt[4]{(N + 1)}\, T$.

> *Note.* This result cannot hold for very large N, because we have ignored the increase in surface area of successive shields.

S146 When a body at absolute temperature T absorbs a quantity of heat ΔQ, the change in entropy $\Delta S \geq \Delta Q / T$. The equality holds when the process is reversible.

Let ΔT_1 and ΔT_2 denote small changes in temperatures of the bodies at temperatures T_1 and T_2, respectively, as a result of receiving quantities of heat ΔQ_1 and ΔQ_2 by a reversible process. The change in the entropy of the whole system is

$$\Delta S_1 + \Delta S_2 = \frac{\Delta Q_1}{T_1} + \frac{\Delta Q_2}{T_2} \geq 0,$$

which we can write as

$$\Delta Q_1 T_2 + \Delta Q_2 T_1 \geq 0.$$

The ratio of the heat transfer to the change of the temperature is the same for both bodies, since their masses are equal. The above equation can therefore be written as

$$(\Delta T_1)T_2 + (\Delta T_2)T_1 = \Delta(T_1 T_2) \geq 0.$$

This means that the geometric mean of the temperatures of the two bodies cannot decrease during the process (though it may increase).

This relationship is valid throughout the process, and hence for the initial and final states, i.e. the common final temperature has to be at least $\sqrt{T_1 T_2}$. If no energy were taken out of the system the common final temperature would be the arithmetic mean of the two initial temperatures. Thus the maximum energy that can be taken out is

$$\left(\frac{T_1 + T_2}{2} - \sqrt{T_1 T_2}\right) mc,$$

where m is the total mass of the water and c is its specific heat.

S147 The number of microstates available to a system of N objects confined to a space of volume V is proportional, both classically and quantum mechanically, to V^N. In the current case, initially there are $2N_A$ molecules of helium occupying $2V_0$ ($V_0 \approx 0.0224 \text{ m}^3$), and correspondingly for the oxygen. After the partition is removed there are $5N_A$ molecules occupying $5V_0$ and the ratio of the new number of microstates to that before is

$$\frac{W_2}{W_1} = \frac{(5V)^{2N_A}}{(2V)^{2N_A}} \times \frac{(5V)^{3N_A}}{(3V)^{3N_A}} = \frac{(5V)^{5N_A}}{(2V)^{2N_A}(3V)^{3N_A}}.$$

Thus the change in entropy is

$$\Delta S = k \ln \frac{W_2}{W_1} = kN_A(5\ln 5 - 2\ln 2 - 3\ln 3) = 27.9 \text{ J K}^{-1}.$$

Note. If we had investigated the mixing of two identical gases (e.g. two moles of helium with three moles of helium) both at s.t.p., the calculated entropy change would seem to be the same. However, it is obvious there can be no entropy change, because physically nothing happens then. This is called the *Gibbs paradox*. The resolution of this paradox depends upon the indistinguishability of identical particles.

S148 Consider the compression stroke of the pump (shown in the figure) in two stages. Initially, both valves are closed and the piston isothermally compresses the air. When the pressure in the pump equals that in the container, the inner valve opens and the total amount of air is isothermally compressed. The moment when the inner valve opens becomes later and later, which makes calculation of the total work done rather complicated. Fortunately, there is a simpler method!

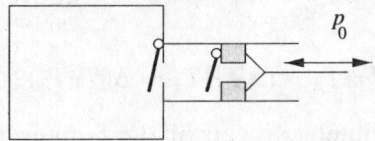

Consider the amount of air that is in the container at the end of the process. This is the 10 litres initially present and a further nine times this amount, i.e. 90 litres (initially at atmospheric pressure). This amount of air occupies a volume of 10 litres.

According to the first law of thermodynamics, the sum of the work W done on the air and the heat Q transferred to it, is equal to the increase in the internal energy of the air. In the present situation, the temperature of the air does not change, and therefore its internal energy is unaltered. Thus, $W + Q = 0$, i.e. the work done on the air equals the heat $-Q$ it gives out.

Further, the change in entropy of the air at constant temperature T, in a reversible process, is $\Delta S = \Delta Q/T$. If the change in entropy of the air can be calculated, the heat given out and the work done can also be determined.

If N molecules are compressed into a ten-times smaller space than they originally occupied, then the number of possible microstates is $1/10^N$ times their original number. According to the statistical interpretation of entropy, the change in entropy of the gas is the logarithm of this number multiplied by the Boltzmann constant,

$$\Delta S = k \ln \left(\frac{1}{10^N} \right) = -Nk \ln 10.$$

This gives

$$-\Delta Q = -T\Delta S = NkT \ln 10 = nRT \ln 10 = 10 p_0 V_0 \ln 10$$

for the heat given out by the gas. The work done is thus

$$W = 10 \times 10^5 \, \text{Pa} \times 10^{-2} \, \text{m}^3 \times \ln 10 \approx 23 \, \text{kJ}.$$

S149 There has to be an electric field between the Earth and the distant planet, as there is a potential difference between them. This electric field causes a charge separation in the wall of the spaceship, and therefore the electric field inside the spaceship is zero (the Faraday cage effect).

The potential difference between the spaceship and the planet changes in the course of the flight, being roughly proportional to the distance from the planet; it is very large at the beginning of the flight but slowly decreases and becomes zero at the end. Thus, when the spaceship lands on the planet, its electric potential is exactly equal to that of the planet and the astronauts can get out safely.

S150 The energy of a capacitor of capacitance C carrying a charge Q is $Q^2/(2C)$. If the change in energy of the capacitor can be found, the change in its capacitance can also be calculated.

The energy of the capacitor is higher when it is dented, since the surface charges have been moved in a direction opposite to that in which their mutual repulsion acts. Further, an electrostatic field $\mathbf{E}$ has an energy $\varepsilon E^2/2$ per unit volume, and an alternative view is that, when the capacitor is dented, the electric field exists in a volume where it was not previously present.

If the surface of the capacitor is only changed a little, the electric field near the surface can be taken as the same as the original one. Thus, the change in energy depends purely on the change in volume and not on the actual shape of the indentation.

Imagine that the original capacitor is hammered so that its volume decreases by 3 per cent, but its shape remains spherical. Its radius is thus reduced by 1 per cent (as the volume of the sphere is proportional to the cube of its radius). The ratio of the energy of such a reduced spherical capacitor to that of the original sphere is the same as the ratio of the energy of the dented capacitor to the original one. Thus, the relative changes in their capacitances are the same as well.

Finally, the capacitance of a spherical capacitor is proportional to its radius. The capacitance of the reduced capacitor is therefore 1 per cent smaller, and the capacitance of the dented capacitor in the original question must have decreased by the same amount.

S151   It can be proved that if equal amounts of charge are carried by F and F^* then the electrostatic energy of the configuration belonging to F is lower.

Let us start from the new surface F^*. Imagine that the charges on it are 'fastened' to the surface, and that the surface is then hammered in such a way as to displace the charges perpendicular to the original surface. The charges have then moved in the direction of the force acting on them. As the surface was originally an equipotential, the field direction was perpendicular to the surface. Thus, the energy of the system decreases in the course of the deformation. (An outward force acts on the surface charges, regardless of their signs – a field directed outwards emanates from the positive charges, while an inwardly directed field is produced by the negative ones.)

The new surface will not be equipotential, but, if the fixed charges are 'released', they migrate, warming the metal up a little whilst doing so. The electrostatic energy of the system is therefore lower in the new equilibrium position.

This process can be repeated until surface F^* is transformed into surface F. The electrostatic energy decreases all the time, while the total charge Q on the metal does not change. Since the electrostatic energy depends on the capacitance, C as $Q^2/2C$, the capacitance of the surface F has to be greater than that of the surface F^*.

S152   The capacitance of the plane capacitor of surface area A is $C = \varepsilon_0(A/d)$. The energy of the capacitor, when connected to a voltage V, can be expressed as

$$W = \frac{1}{2}CV^2 = \varepsilon_0 \frac{AV^2}{2d}.$$

When the distance between the plates is increased from d to $2d$, both the capacitance and the energy of the capacitor decrease to half of their original values. This may be a surprising result, as pulling apart the plates of the capacitor requires positive work. However, when charge ΔQ leaves the plates of the capacitor it increases the energy of the battery (charges the battery) by $\Delta Q V$. Since $\Delta Q = \Delta C V$, the increase in the energy of the battery is exactly twice the decrease in the energy of the capacitor. The increase in the energy of the battery is due half to the decrease in energy of the capacitor and half to the work done in pulling the plates apart.

Note. The above statement can be verified by direct calculation of the work done in pulling the plates apart. The force of attraction acting between the plates of a plane capacitor can be calculated using the relation $F = Q^2/(2\varepsilon_0 A)$. Substituting for the charge in terms of voltage and capacitance gives

$$F = \frac{C^2 V^2}{2\varepsilon_0 A} = \frac{\varepsilon_0 A V^2}{2d^2}.$$

If the distance between the plates is denoted by x, the work done can be calculated as

$$W = \int_d^{2d} F(x)\, \mathrm{d}x = \frac{\varepsilon_0 A V^2}{2} \int_d^{2d} \frac{\mathrm{d}x}{x^2} = \frac{\varepsilon_0 A V^2}{2}\, \frac{1}{2d} = \frac{\varepsilon_0 A V^2}{4d}.$$

So the work done in pulling the plates apart is equal to the decrease in energy of the capacitor, and these two quantities together increase the energy of the battery.

S153 Imagine that the current I_0 flows in a superconducting, short-circuited coil of N turns, with cross-section $A = \pi R^2$ and length x_0. The magnetic field strength inside the coil is then $B = \mu_0 I N / x_0$ and the total magnetic flux $\Phi = BAN = \mu_0 I_0 N^2 A / x_0$. Even if the length of the coil changed for some reason, the magnetic flux would not change, as to do so would induce a voltage and hence an 'infinitely' large current in the coil of zero resistance. Thus the current must vary with the length x as $I(x) = I_0 x / x_0$. The inductance of a coil of length x is $L(x) = \mu_0 N^2 A / x$, and the magnetic energy of a coil with current I flowing through it is

$$W_\mathrm{m} = \frac{1}{2} L I^2 = \mu_0 \frac{N^2 I_0^2 A}{2x_0^2}\, x.$$

Thus the energy of the coil is proportional to x, i.e. $W_\mathrm{m}(x) \propto x$. The proportionality constant F_0 is the magnetic force of contraction, since work $W = F_0 x$ is needed to re-stretch the coil by length x.

The coil is in equilibrium when the magnetic force of contraction balances the elastic force $F(x) = k(x_0 - x)$, i.e. when the change of spring length is

$$\Delta x = x_0 - x = \frac{F_0}{k} \approx \mu_0 \frac{\pi I_0^2 N^2 R^2}{2kx_0^2}.$$

S154 (i) The horizontal force (measured in the direction of increasing x, i.e. to the left in the figure in the problem) exerted on magnet A is

$$F(x) = -\frac{K}{x^n} + mg\frac{(d+s-x)}{\ell}, \tag{1}$$

where $d + s$ is the distance from magnet B to the unperturbed position of magnet A. The condition of equilibrium is

$$F(x)|_{x=d} = 0. \tag{2}$$

The stability of equilibrium depends on the behaviour of the function $F(x)$ close to $x = d$. The function can be monotonically increasing or decreasing (Fig. S154.1).

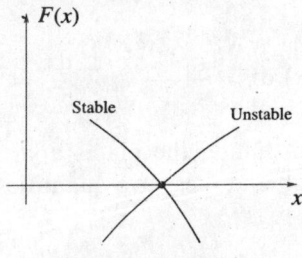

Fig. S154.1

For stable equilibrium, its derivative has to be negative. In the unstable case the derivative is positive, and in the limiting case of neutral equilibrium it is zero, i.e.

$$F'(x)|_{x=d} = 0. \tag{3}$$

Using the expression in (1) for the force $F(x)$, condition (3) can be formulated as

$$\frac{nK}{d^{n+1}} - \frac{mg}{\ell} = 0. \tag{4}$$

But condition (2) can be written in the form

$$-\frac{K}{d^n} + \frac{mg}{\ell}s = 0. \tag{5}$$

From these equations we get

$$n = \frac{d}{s} = 4 \quad \text{and} \quad K = mgd^4\frac{s}{\ell}. \tag{6}$$

(ii) In the vertical tube (Fig. S154.2) the repulsive force acting on magnet A is $F_{\text{vert}}(x) = +K/x^4 - mg$.

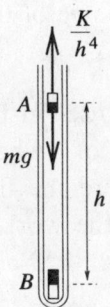

Fig. S154.2

Equilibrium levitation occurs with the magnets a distance h apart, where

$$F_{\text{vert}}(x)|_{x=h} = 0.$$

Taking into account result (6) we get

$$h = \left(\frac{K}{mg}\right)^{1/4} = d\left(\frac{s}{\ell}\right)^{1/4} = 4 \text{ cm} \times \left(\frac{1}{100}\right)^{1/4} \approx 1.3 \text{ cm}.$$

S155 The energy ultimately stored in the capacitor is $\frac{1}{2}CN^2\mathscr{E}^2$ in both cases. Direct connection results in an equal amount of wasted energy. For charging by stages, the total work done by the battery is

$$\Delta Q \sum(\Delta V) = C\mathscr{E}(\mathscr{E} + 2\mathscr{E} + \cdots + N\mathscr{E}) = \frac{1}{2}C\mathscr{E}^2 N(N+1).$$

The wasted energy is thus

$$\frac{1}{2}C\mathscr{E}^2[N(N+1) - N^2] = \frac{1}{2}C\mathscr{E}^2 N,$$

i.e. only $1/N$ of the original loss.

Note. You can obtain the same result by considering the following graph.

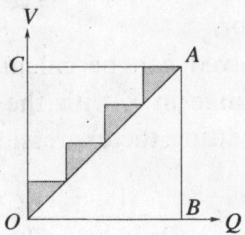

The triangle OAB represents the stored energy of the capacitor, and the triangle OAC corresponds to the energy wasted by charging it in a single step. The small shaded triangles show the energy wasted when using multiple (e.g. $N = 4$) steps. The total area of the shaded triangles is $1/N$ of that of OAC.

S156 The stored energy is $Q^2/2C$ in both cases and $C_{oil} = \varepsilon C_{air}$. Thus the energy is increased by a factor of ε when the oil is poured out. The catch is that the oil is attracted to where the field is strongest and work has to be done to extract it from there; the work needed is at least as great as the increase in stored electrical energy.

S157 (i) Let the distance between the capacitor plates of area A be d, and the charge on them be Q. Examine the situation when a length x of the insulating sheet has already been slid between the plates of length ℓ as shown in the figure.

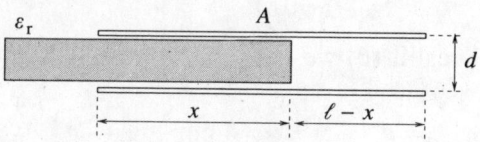

The capacitance of the plane capacitor can be calculated as that of two capacitors connected in parallel, one filled with dielectric and the other empty:

$$C = \varepsilon_0 \varepsilon_r \frac{A}{d} \frac{x}{\ell} + \varepsilon_0 \frac{A}{d} \frac{\ell - x}{\ell} = \frac{\varepsilon_0 A}{\ell d} [(\varepsilon_r - 1)x + \ell].$$

The total capacitance can be seen to increase linearly with x from $\varepsilon_0 A/d$ to ε_r times its initial value. As the charge is constant in the present case the energy of the capacitor W_{cap} can be written as

$$W_{cap} = \frac{Q^2}{2C} = \frac{Q^2 \ell d}{2\varepsilon_0 A [(\varepsilon_r - 1)x + \ell]}.$$

It is clear that the energy of the capacitor decreases as x increases! This means that the work done in sliding the insulator in is negative, i.e. the plates effectively suck in the insulator.

The size of the force involved can be calculated by equating the work corresponding to a small change in x with the resulting change in energy ($F\Delta x = \Delta W_{cap}$), and differentiating the expression for W,

$$F = \frac{dW}{dx} = -\frac{(\varepsilon_r - 1)Q^2 \ell d}{2\varepsilon_0 A [(\varepsilon_r - 1)x + \ell]^2}.$$

The magnitude of the force is greatest at $x = 0$, since the denominator in the above expression increases with x. It is easy to show that when $x = \ell$, the magnitude of the force has decreased by a factor ε_r^2.

> *Note.* How does the electric field exert a force on the dielectric parallel to the plates? If the electric field between the plates were homogeneous and perpendicular to the plates, and zero outside them (i.e. the usual picture of capacitors), then *no* force could act on the insulating sheet. The phenomenon is explained by the curvature of the electric field that is inevitably present at the edge of the plates.

(ii) The force acting on the dielectric cannot depend on the interaction of the capacitor with its surroundings. Therefore the previous result would be valid in the constant voltage case if it were not for the fact that the charge on the capacitor changes in accordance with $Q = CV$ as soon as the insulating sheet is placed between the plates. Substituting for the charge (which depends on x) into the above expression gives

$$F = -\frac{\varepsilon_0 A (\varepsilon_r - 1)\, V^2}{2\ell d},$$

i.e. in this case the force acting on the insulating sheet is constant.

An interesting conclusion can be drawn from the expression for the energy of the capacitor at constant voltage,

$$W_{\text{cap}} = \frac{CV^2}{2} = \frac{\varepsilon_0 A\, [(\varepsilon_r - 1)x + \ell]\, V^2}{2\ell d}.$$

This shows that, in this case, the energy of the capacitor increases linearly with x, and that the change of energy ΔW_{cap} corresponding to a small displacement Δx of the insulator is

$$\frac{\Delta W_{\text{cap}}}{\Delta x} = \frac{\varepsilon_0 A(\varepsilon_r - 1)V^2}{2\ell d}.$$

Apart from the sign, this formula is exactly the same as that for the force acting on the insulating sheet. These results can be summarised in the following way: when the dielectric is slid between the plates, the system does work on it (i.e. pulls it in), while the energy of the capacitor *increases* by the same amount. This is possible because the energy of the battery decreases by twice this amount during the process. The decrease in stored battery energy occurs because the capacitance of the capacitor (and therefore its charge) increases, and the battery has to supply the additional charge. The calculation of the work done by the battery is left to you the reader.

S158 Number the resistors, starting with the last element in the chain. As a current of 1 A flows through the first resistor, a current of 1 A has to flow through the second one as well, thus there is a potential difference (p.d.) of 1 V across each resistor. As a consequence, the p.d. across the third resistor is $(1 + 1) = 2$ V, and the current flowing through it must be 2 A. The current flowing through the next resistor is $(1 + 2) = 3$ A. The current in the fifth resistor can be determined using the p.d. $(2 + 3) = 5$ V across the resistors with currents of 2 and 3 A, respectively, flowing through them, and so on, as shown in Fig. S158.1.

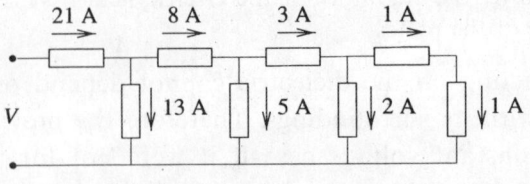

Fig. S158.1

Consider the chain of resistors to be built starting with the last element and then connecting the consecutive elements, alternately in series and parallell, throughout the chain. The sum of the currents flowing through the two previous resistors flows through the following series resistor (Kirchhoff's first law). The next element connected in parallel creates a new loop in the chain, and therefore the p.d. across this resistor equals the sum of the p.d.s across the two previous ones (Kirchhoff's second law). Since the numerical values of the p.d. and the current are identical for a 1-Ω resistor, the sum of the currents of the two previous resistors is the same as the current of the new resistor connected in parallel. Thus, in this so-called ladder circuit, Kirchhoff's laws are satisfied in such a way that the current flowing through each resistor (and the p.d. across it) is equal to the sum of the corresponding quantities for the two previous elements.

Notice that the numerical values of the currents (or p.d.s) are the terms of the Fibonacci series: 1, 1, 2, 3, 5, 8, 13, 21. The p.d. across the two last resistors is $(21 + 13) = 34$ V, and this must also be the p.d. across the circuit input. As a total of 21 A flow as the result of applying a p.d. of 34 V, the equivalent resistance of the circuit is of $34/21 = 1.619\,05\ \Omega$.

If one more element is connected to the chain (in parallel) then the p.d. across it is unchanged at 34 V, but the total current increases to $(21 + 34) = 55$ A. In this case the equivalent resistance is $34/55 = 0.618\,18\ \Omega$. If yet another element is connected to the circuit, a current of 55 A flows through it and the input p.d. increases to $(34 + 55) = 89$ V. The total resistance of the chain is then $89/55 = 1.618\,18\ \Omega$.

If the ladder circuit is extended further and further, an 'infinite' chain is obtained. The equivalent resistance of this chain can be calculated using the fact that adding two more elements does not change its resistance. Thus, the whole chain can be replaced by a single resistor of resistance R, which is such that if two 1-Ω resistors are connected to it, one in parallel and the other in series with the combination, the equivalent resistance of the new circuit will also be R (*see* Fig. S158.2).

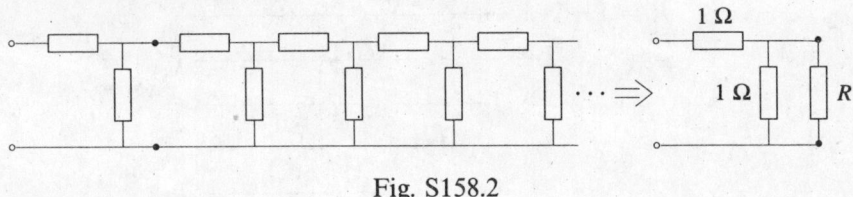

Fig. S158.2

The condition for this is

$$R = 1\ \Omega + \frac{1}{(1/1\ \Omega) + (1/R)},$$

which yields the following quadratic equation for the numerical value of R,

$$R^2 - R - 1 = 0.$$

The positive root of this equation gives the equivalent resistance for an 'infinite' chain as,

$$R = \frac{1 + \sqrt{5}}{2} \approx 1.618\,03\ \Omega.$$

We see that the equivalent resistance of a chain of eight–ten elements approximates very well that of the 'infinite' chain. Hence a ladder circuit with relatively few elements can be considered as infinite.

> *Note.* (i) In practice, the neutral wires of overhead electric supply networks can be considered as ladder circuits; the neutral wires are fastened to poles and earthed at, say, every tenth pole. Such a ladder circuit consists of two types of resistors, but the equivalent resistance of the 'infinite' chain can be calculated using the above method.
>
> (ii) It is of interest to note that the above quadratic equation is the golden ratio equation, the solution of which is the golden mean, $(1 + \sqrt{5})/2 = 1.618\,03\ldots$. As shown, this is the same as the numerical value of the equivalent resistance of the infinite ladder circuit. Furthermore, the ratio of consecutive elements of the Fibonacci series was shown to approach the golden mean surprisingly quickly. It is also easy to prove that dividing the even elements by the previous odd ones, the golden mean is approached from below, while dividing the odd elements by the previous even ones, it is approached from above.

Finally, for the sake of the aesthetic pleasure produced by multiple-level fractions, it is worth expressing with 1s the equivalent resistance of the infinite chain made up of 1-Ω resistors. This time, the elements are considered in order, not from the end but from the beginning, i.e. starting at the left hand end of the chain:

$$R = 1 + \cfrac{1}{1 + \cfrac{1}{1 + \cfrac{1}{1 + \cfrac{1}{1 + \cfrac{1}{1 + \cfrac{1}{1 + \cdots}}}}}}$$

If the $\cdots$ at the end of the formula is replaced by a 1, then the equivalent resistance of the original eight-element chain is obtained.

S159   We discuss first the grid of ohmic resistors and reason as follows. Consider first a grid point at which a current I enters and then flows away 'to infinity'. From symmetry, identical currents of magnitude $I/4$ will start from this point and travel along the four possible directions as shown in Fig. S159.1.

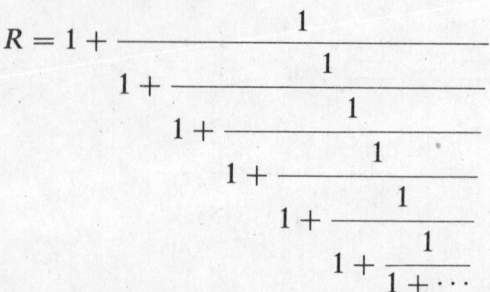

Fig. S159.1

Now consider the neighbouring grid point and let current I flow out of it (independently of the previous reasoning). Again, identical currents of magnitude $\frac{1}{4}I$ flow through the four identical resistors adjacent to the point (*see* Fig. S159.2).

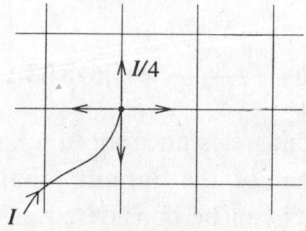

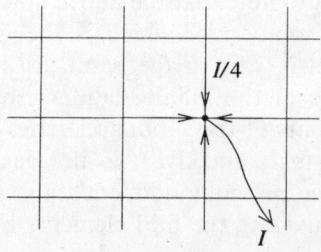

Fig. S159.2

Now consider superposing these two cases. Because of the linearity of the circuit equations, scalar quantities (e.g. currents and potentials) simply add. A voltage (p.d.) V appears between the two neighbouring grid points, current I goes into one of them and current I flows out of the other. The currents flowing elsewhere in the grid cannot be easily determined, but the current flowing between these two neighbouring points can. The two currents discussed above add in this resistor, i.e. a total current of $\frac{1}{2}I$ flows through it (*see* Fig. S159.3).

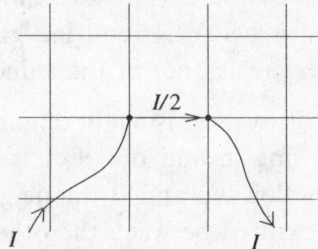

Fig. S159.3

But if current $\frac{1}{2}I$ flows through a resistance R then a p.d. of $V = RI/2$ appears across it. The equivalent resistance between the two neighbouring points is therefore $R_e = V/I = R/2$.

The equivalent values for infinite grids of capacitors and self-inducting coils can be calculated similarly as $C_e = 2C$ and $L_e = \frac{1}{2}L$.

S160 Denote the number of grid points of the polyhedron by c and the number of lines meeting at a grid point by n (e.g. in the case of a dodecahedron, $c = 20$ and $n = 3$). If a current of 1 amp flows in at one grid point and $1/(c-1)$ amps are taken out at all other grid points, then, by symmetry, the current flowing in each of the lines which meet at the point where the current enters is $I = 1/n$ amps.

Superimpose on the previous current distribution, that produced by a current of 1 amp flowing out of the neighbouring grid point and currents of $1/(c-1)$ amps flowing into all the other points (including the original point). The current of $I = 2/n$ amps flowing through the 'direct' resistor (i.e. the one directly joining the two neighbouring grid points) causes a p.d. of $V = 2/n$ volts to appear across it. Since the current flowing through the whole circuit is $I = 1 + 1/(c-1) = c/(c-1)$ amps, the equivalent resistance is

$$R_e = 2(c-1)/nc.$$

This figure is $19/30 \ \Omega$ for the dodecahedron and $1/2 \ \Omega$ for the infinite square grid. It can be calculated similarly for any regular shape.

S161   Let us denote the equivalent resistances of the original grid by R_e^{orig} and of the truncated circuit by R_e^{trun} We can consider the original grid as a circuit of two resistors connected in parallel; the two resistors are R_e^{trun} and R, the latter being the removed resistor. It is then easy to write an equation which involves R_e^{trun}, namely,

$$R_e^{orig} = \frac{R \times R_e^{trun}}{R + R_e^{trun}} \quad \text{or} \quad R_e^{trun} = \frac{R \times R_e^{orig}}{R - R_e^{orig}}.$$

For example, in the case of the infinite two-dimensional square grid the equivalent resistance between two neighbouring grid points is $R_e^{orig} = R/2$. It follows that the equivalent resistance of the truncated circuit is $R_e^{trun} = R$.

S162   The 'trick' of superposition is again employed to combine separate discussions of currents flowing in and out. Let A denote the corner of the square where the current I_0 flows in and B the neighbouring corner where it flows out. The p.d. V is measured between the other two corners (C and D) as shown in the figure.

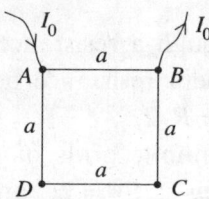

If the current I_0 is introduced at point A (and flows towards points at zero potential infinitely far away), it is distributed (hemi)spherically symmetrically in the half-space containing the matter, i.e. at a distance r from point A, the current density $j(r)$ is $j(r) = I_0/2\pi r^2$. This relation is not valid for very small values of r, i.e. when r is not much greater than the size of the electrode at A.

The local version of Ohm's law is the differential law that expresses the current density $\mathbf{j}$ in terms of the electric field strength $\mathbf{E}$ and the local resistivity ρ, namely $\mathbf{j}(r) = \mathbf{E}(r)/\rho$.

The magnitude of the electric field strength in the half-space can be determined using this relation as

$$E(r) = \frac{I_0\rho}{2\pi r^2}.$$

The potential function (and hence the p.d.) could be determined from the field strength by integration. In this case, however, it can be obtained using a simple analogy. The electric field of a point charge Q is inversely proportional to r^2, its potential is inversely proportional to r and the proportionality coefficient is the same in both cases ($E = kQ/r^2$ and $V = kQ/r$, respectively). This means that the potential field corresponding to the electric field strength determined above is $V = I_0\rho/(2\pi r)$.

The nearer a point is to the electrode where the current flows in, the higher its potential. The potential $V_D = I_0\rho/(2\pi a)$ of point D is therefore higher than $V_C = I_0\rho/(2\sqrt{2}\pi a)$. The p.d. between the two points is $I_0\rho(2 - \sqrt{2})/(4\pi a)$.

We next discuss the flow of the current I_0 out through point B. Everything is the same as in the previous case, except that the signs of the quantities (current, current density, field strength and potential) are reversed. The potential in the half-space under investigation is described by the function $V = -I_0\rho/(2\pi r')$, where r' is the distance from point B. The potential of point C is lower than that of point D, i.e. point D is again more positive than C. The p.d. between the two points is the same as previously.

If the two previous cases are now superposed we return to the original problem, and the p.d. between points C and D is exactly twice either of the above p.d.s, i.e. $V_0 = I_0\rho(2 - \sqrt{2})/(2\pi a)$. Apart from given data, this expression contains only the resistivity ρ. Therefore the solution of the problem is

$$\rho = (2 + \sqrt{2})\pi a V_0/I_0.$$

Note. This method is widely used in 'real life', e.g. to determine the average resistivity of rocks. The measurements are, naturally, not made on infinite half-spaces, but on volumes and planes with linear sizes much larger than the side a of the square.

S163 First connect the battery to the terminals of the resistor through the ammeter, and then connect the voltmeter across the same terminals, as shown in Fig. S163.1.

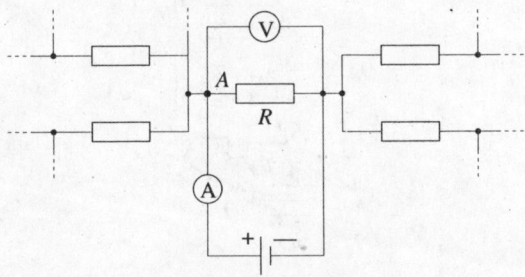

Fig. S163.1

It is naïve to assume that the quotient of the measured potential difference and current gives the resistance, because we cannot be sure that at junction

A all the current coming from the ammeter flows through the resistor *R*, and that some does not flow towards other parts of the circuit.

The problem can be readily solved with the help of short circuits. Using zero-resistance wires we connect all of *A*'s neighbouring junctions to the same terminal of the battery as that to which the ammeter is connected, as shown in Fig. S163.2.

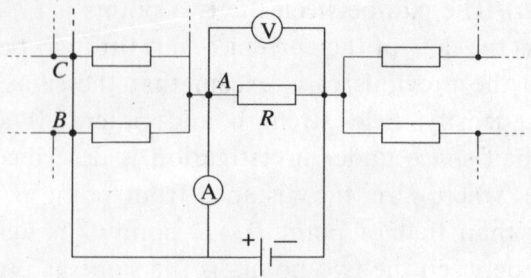

Fig. S163.2

As the internal resistance of the ammeter is negligible, the junctions $(B, C, D, \ldots)$ and *A* are equipotential points. Consequently, there is no current flowing between them and the current from the ammeter must all flow through the resistor *R*. It is possible that the battery has to provide additional currents flowing towards junctions $B, C, D, \ldots$, but these have no influence on our measurement.

> *Note.* If resistor *R* is connected in parallel with other resistors, then there is no way to measure the resistance of these resistors separately – only their equivalent resistance can be found.

S164   Let current *I* flow into the cube at one point and flow out at the point diagonally opposite to it (*see* Fig. S164.1).

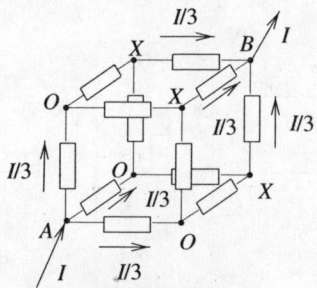

Fig. S164.1

Symmetry prescribes that the current flowing through the three resistors that meet at each endpoint of the diagonal is $I/3$, and hence that the voltages across these resistors are identical. Thus, the sets of points denoted by *O* and

X in the figure are each equipotentials, i.e. the members of each set can be joined together without disturbing the system.

When the equipotential points are joined, the circuit can be redrawn as shown in Fig. S164.2, and the equivalent resistance can be calculated mentally. If all the resistors have resistances of 1 Ω, then the equivalent resistance is 5/6 Ω.

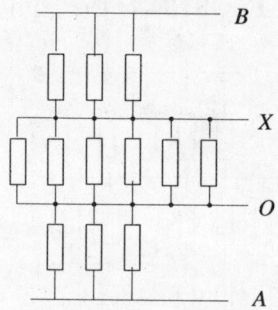

Fig. S164.2

A one-dimensional 'cube' is simply a straight section, its resultant resistance is itself, i.e. 1 Ω. The two-dimensional 'cube' is the square. Two resistors emanate from one end of a diagonal and two resistors converge at the opposite end of the same diagonal. If current flows through the square, the endpoints of the other diagonal are equipotential. Therefore two lots of two parallel resistors are connected in series. For a square the equivalent resistance is again 1 Ω. As we have already seen, for a three-dimensional cube, three resistors start from each end of the diagonal and the remaining six resistors join equipotential surfaces.

A four-dimensional 'cube' can be obtained by a parallel displacement of a three-dimensional cube in the direction of the fourth dimension, followed by the joining of corresponding points. The four-dimensional 'cube' therefore has $12 + 12 + 8 = 32$ edges (12 for each normal cube plus eight to join corresponding corners). A distorted projection of such a 'cube' can be made for investigative purposes, as shown in Fig. S164.3, replacing the displacement by a magnification. Figure S164.3 shows one diagonal, AB, from the eight possible ones.

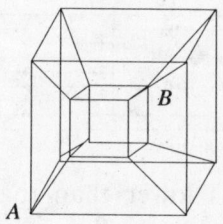

Fig. S164.3

For a four-dimensional 'cube', four lines lead out of any one vertex and the endpoints of these are equipotentials. The same is true at the other end of the diagonal. The shortest way from one of these equipotential surfaces to the other is through any two resistors of the remaining 24. This means that the 'inner' points of the 24 resistors are also equipotential, i.e. two lots of 12 resistors are connected in parallel and these are then connected in series. The redrawn circuit is shown in Fig. S164.4; the equivalent resistance is 2/3 Ω.

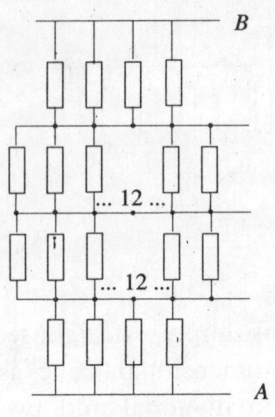

Fig. S164.4

Note. The problem can be generalised to n-dimensional 'cubes' if, after connecting equipotential points, the circuit can be decomposed into sets of identical resistors connected in parallel with the sets then connected in series. In general, the equivalent resistance across a diagonal of an n-dimensional 'cube' made of 1-Ω resistors is R_n Ω, where R_n is given by

$$R_n = \frac{1}{1\binom{n}{1}} + \frac{1}{2\binom{n}{2}} + \frac{1}{3\binom{n}{3}} + \cdots + \frac{1}{n\binom{n}{n}}.$$

S165   Let I denote the current flowing in the wire, A the cross-section of the wire, and ρ_1 and ρ_2 the resistivities of the metals. Ohm's law for a wire of length ℓ gives $V = I\rho\ell/A$, which yields $E = V/\ell = \rho I/A$ for the electric field strength in the wire.

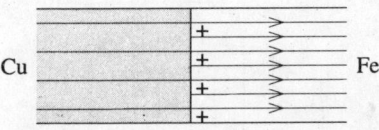

The resistivity of copper is lower than that of iron, and therefore the electric field strength has to be smaller in the copper than in the iron.

According to Gauss's law, the difference in the electric field strengths implies an accumulation of charge at the boundary of the two metals (*see figure*). The net accumulated charge is

$$Q = \varepsilon_0 A(E_{Fe} - E_{Cu}) = \varepsilon_0 I(\rho_{Fe} - \rho_{Cu}).$$

It is interesting that this quantity depends purely on the current and material constants, and not on the cross-section of the wire.

Substituting the known data, the charge is found to be $Q \approx 5 \times 10^{-21}$ C, which is only $\frac{1}{30}$ of an elementary charge! Though a measurable macroscopic current flows through the wire, the accumulated charge is only a small proportion of the microscopic elementary charge. This strange result shows that classical electrodynamics (imagining charge carriers as small balls) cannot always correctly describe electrical phenomena. Only the application of the more sophisticated laws of quantum theory and statistical physics can give an accurate description.

S166 In SI units the jet's speed is 200 m s^{-1}.

(i) The magnetic field is vertically downwards, and the induced voltage is $80 \times 6 \times 10^{-5} \times 200 = 960$ mV, with the starboard (right) wing tip at the higher potential.

(ii) The motion is parallel to the field, and so no potential difference is developed.

(iii) Since the field is a dipole its strength at the Equator is one-half that at the Pole. The field is horizontal, and the induced voltage is $8 \times 3 \times 10^{-5} \times 200 = 48$ mV, with the bottom of the jet at the higher potential.

(iv) The vertical field is $5 \times 10^{-5} \times \sin 66°$ T and the northward component of the velocity is $200/\sqrt{2}$ m s^{-1}, leading to a 520 mV potential difference across the wings with the starboard tip at the higher potential. In the same way, a p.d. of 23 mV appears between the top (higher potential) and the bottom of the jet's body.

S167 Let that the rod move with speed v and acceleration a along the inclined plane, while current I flows in it. The magnetic field brakes the rod in accordance with Lenz's law, and its equation of motion is

$$ma = mg \sin \alpha - B\ell I.$$

This equation is the same in all three cases. The differences result from the different relationships between the induced voltage and the current flowing in the rod.

(i) If the circuit is closed by an ohmic resistor R, the current I and the induced voltage $V = B\ell v$ are connected by the relationship $I = V/R =$

$B\ell v/R$. This shows that the braking force increases in proportion to the speed, with the result that the rod experiences decreasing acceleration and ultimately travels with uniform speed. This final maximum speed $v_{\max}$ can be found from the equation of motion by setting $a = 0$,

$$v_{\max} = \frac{mgR\sin\alpha}{B^2\ell^2}.$$

(ii) If the circuit is closed by a capacitor of capacitance C, the relationship between the induced voltage and the current is different. The charge on the capacitor is determined by the induced voltage, and given by

$$Q = CV = CB\ell v.$$

Note that the current flowing through the rod is equal to the time derivative of this, i.e.

$$I = \frac{dQ}{dt} = C\frac{dV}{dt} = CB\ell\frac{dv}{dt} = CB\ell a.$$

In other words, the current flowing in the rod is directly proportional to the acceleration of the rod. If the above expression for the current is substituted into the equation of motion, the rod is found to move on the rails with uniform acceleration

$$a = \frac{mg\sin\alpha}{m + B^2\ell^2C}.$$

Induction decreases the acceleration caused by gravity by, in effect, increasing the inertial mass of the rod. The speed of the rod and the charge on the capacitor are both directly proportional to the time elapsed.

(iii) If the circuit is closed by a coil of inductance L, the relationship between the induced voltage and the current is

$$L\frac{dI}{dt} = B\ell v = B\ell\frac{dx}{dt}.$$

We note that, since $I = 0$ and $x = 0$ at the start of the motion, the above formula implies that the current is proportional to the x-coordinate, $LI = B\ell x$. Substituting for the current, from this relationship into the equation of motion, gives

$$ma = mg\sin\alpha - \frac{B^2\ell^2}{L}x.$$

The force acting on the rod is therefore the sum of a constant term and a negative term proportional to the displacement. This is the same as the equation of motion of a body hung on a spring and then released. Thus, the rod makes harmonic oscillations about an equilibrium position

$$x_0 = \frac{mgL \sin \alpha}{B^2 \ell^2}.$$

The amplitude of the oscillation is $A = x_0$, and the dependence of the displacement of the rod on time is

$$x(t) = A(1 - \cos \omega t),$$

where $\omega^2 = \dfrac{B^2 \ell^2}{mL}$.

S168 (i) At the instant when the capacitor is connected, a current $I = V_0/R$ starts flowing in the rod, which experiences a force $F = B\ell I$ and an initial acceleration $a = B\ell V_0/mR$. In accordance with Lenz's law, the voltage induced in the moving rod causes the current flowing in the rod to decrease. The charge Q on the capacitor decreases and consequently so does the voltage across it. Meanwhile the voltage induced in the rod increases, until the two voltages cancel each other out. The rod then continues with its maximum velocity given by

$$B\ell v_{max} = \frac{Q_{min}}{C}. \tag{1}$$

The equation of motion of the rod is

$$m\frac{dv}{dt} = ma = B\ell I = -B\ell \frac{dQ}{dt}, \tag{2}$$

where the acceleration and the current have been expressed as the rates of change in velocity and charge, respectively. The proportionality between the two rates of change holds throughout. The speed of the rod increases from zero to v_{max}, whilst the charge on the capacitor decreases from $Q_0 = CV_0$ to Q_{min}. Equation (2) can therefore be rewritten as

$$mv_{max} = B\ell(Q_0 - Q_{min}).$$

The final velocity and the residual charge on the capacitor can be calculated using equations (1) and (2),

$$v_{max} = \frac{B\ell CV_0}{m + B^2 \ell^2 C} \quad \text{and} \quad Q_{min} = \frac{B^2 \ell^2 C^2 V_0}{m + B^2 \ell^2 C}.$$

(ii) The above relations show that the maximum velocity of the rod is proportional to the initial voltage V_0 across the capacitor. Thus, the final kinetic energy of the rod is proportional to V_0^2 (for given values of C and m), i.e. proportional to the initial energy of the system. The coefficient of proportionality can be regarded as the efficiency η of the apparatus (considered as an electromagnetic gun), and can be written in the form

$$\eta = \frac{\frac{1}{2}mv_{\max}^2}{\frac{1}{2}CV_0^2} \doteq \frac{1}{\left(\dfrac{\sqrt{m}}{B\ell\sqrt{C}} + \dfrac{B\ell\sqrt{C}}{\sqrt{m}}\right)^2}.$$

The product of the two terms in the brackets is 1, and from the inequality between arithmetic and geometric means, it follows that their sum is at least 2. This means that the efficiency of the electromagnetic gun cannot be more than 25 per cent.

> *Note.* If the condition for maximum efficiency $m = CB^2\ell^2$ is substituted into the expression for the final charge on the capacitor we find that $Q_{\min} = V_0C/2$, i.e. only half of the initial charge on the capacitor is left. Thus, only one-quarter of the initial energy of the capacitor is left; one-quarter of it is transformed into the kinetic energy of the rod, and the other half is dissipated in the rod as Joule heat.

S169   (i) The rate of increase of magnetic energy ($E_{\text{magn}} = LI^2/2$) is the difference between the power output of the battery and the power dissipated in the resistor,

$$\frac{dE_{\text{magn}}}{dt} = VI - RI^2 = -R\left(I - \frac{V}{2R}\right)^2 + \frac{V^2}{4R} \le \frac{V^2}{4R}.$$

It is clear that the rate of increase is maximal when $I = V/(2R)$.

(ii) After the switch has been closed, we can write Kirchhoff's law for the circuit as

$$V = IR + L\frac{dI}{dt},$$

which gives the current–time relationship (*see* Fig. S169.1)

$$I = \frac{V}{R}\left(1 - e^{-\frac{R}{L}t}\right).$$

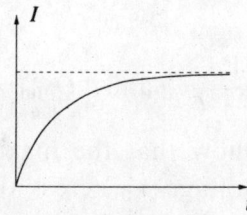

Fig. S169.1

The power dissipated in the resistor by Joule heating is

$$P = RI^2 = \frac{V^2}{R}\left(1 - e^{-\frac{R}{L}t}\right)^2.$$

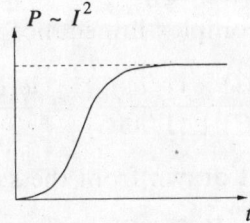

Fig. S169.2

Figure 169.2 is a sketch of the power–time graph, and shows that the power increases monotonically. Whilst the rate of change of power initially increases, it later reaches a maximum, and beyond this point decreases monotonically to zero as shown in Fig. S169.3.

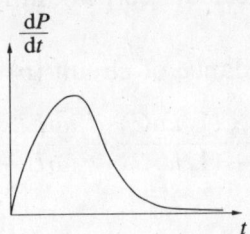

Fig. S169.3

The fastest rate of change of energy occurs, for both the inductor and resistor, when I^2 is changing most rapidly. In part (i) we found that this happens when $I = V/2R$. Substituting this value for I into the above expression for the power, shows that it occurs when $e^{-Rt/L} = \frac{1}{2}$, and hence that $t = (L/R)\ln 2 \approx 0.69\,(L/R)$.

> *Note.* (i) The rate of energy loss from the battery is VI, and so is proportional to the current, which increases monotonically.
>
> (ii) The fastest increase in the current takes place at $t = 0$, but that in I^2 (which is proportional to the magnetic energy of the coil) occurs later.
>
> (iii) Just for fun you may wish to solve the twin of this problem, in which the inductor is replaced by a capacitor.

S170 (i) These circuits are tricky to analyse using differential equations, but become straightforward if complex impedances are employed. The impedance of an inductor of inductance L at angular frequency ω is $i\omega L$, whilst that of a capacitor of capacitance C is $1/i\omega C$. Here i is the square root of -1, i.e. $i^2 = -1$. Combining impedances is governed by the same rules as those which apply to resistances R_1 and R_2 in series $[R = R_1 + R_2]$ and in parallel $[R = R_1 R_2/(R_1 + R_2)]$.

Circuit (a) thus has total complex impedance

$$Z = \frac{i\omega L \times i\omega L}{i\omega L + i\omega L} + \frac{(1/i\omega C) \times (1/i\omega C)}{(1/i\omega C) + (1/i\omega C)} = \frac{i\omega L}{2} + \frac{1}{2i\omega C} = \frac{1 - \omega^2 LC}{2i\omega C}.$$

The magnitude of the current drawn from the source is therefore

$$|I| = \frac{|V_0|}{|Z|} = \frac{V_0|2i\omega C|}{|1 - \omega^2 LC|} = CV_0\omega_0\frac{2x}{|1 - x^2|},$$

when LC is written as ω_0^{-2} and ω/ω_0 is written as x. This is plotted in Fig. S170.1(a). It will be seen that, theoretically, the current increases without limit as ω approaches ω_0; in practice the source will be unable to supply such a current. In any case, real inductors and connectors have non-zero resistance, and the calculated peak in Fig. S170.1(a) is then of finite amplitude.

In a similar way the impedance of circuit (b) is

$$Z = \frac{i\omega L \times (1/i\omega C)}{i\omega L + (1/i\omega C)} + \frac{i\omega L \times (1/i\omega C)}{i\omega L + (1/i\omega C)},$$

leading to

$$|Z| = \frac{2\omega L}{|1 - \omega^2 LC|} \qquad \text{and} \qquad |I| = \frac{CV_0\omega_0|1 - x^2|}{2x}.$$

This is plotted in Fig. S170.1(b). It will be noticed that at $\omega = \omega_0$ the circuit has an infinite impedance and no current is drawn. It is also of interest that the x-dependences in Fig. S170.1(a) and (b) are reciprocal functions.

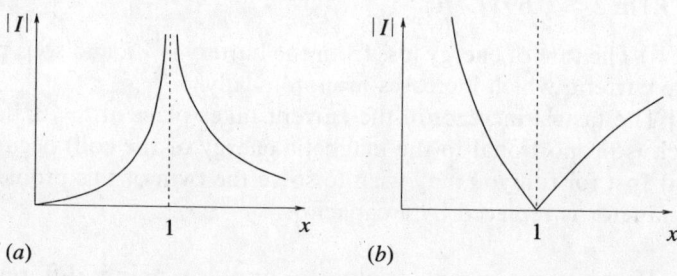

Fig. S170.1

(ii) Inductances of $L, 2L, L/2$, and capacitances of $C, 2C, C/2$, constructed from identical components in series or parallel as needed, are used to make circuits resonant at $\omega_0, \omega_0/2, 2\omega_0, \omega_0/\sqrt{2}, \sqrt{2}\omega_0$. As shown in Fig. S170.2, for two of these there are two alternatives.

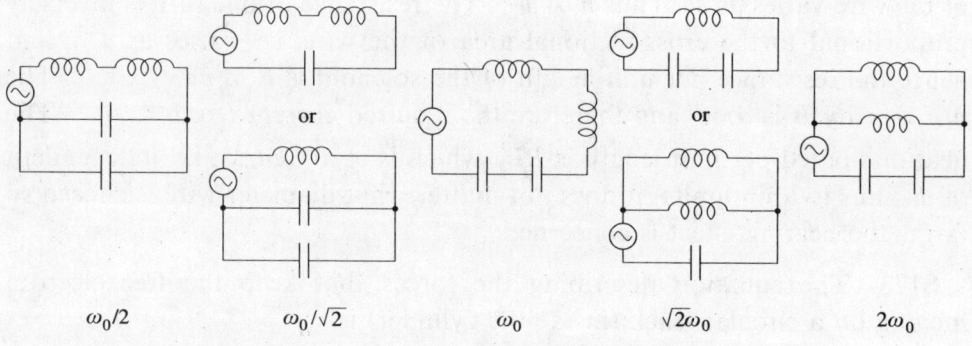

$\omega_0/2$ $\omega_0/\sqrt{2}$ ω_0 $\sqrt{2}\omega_0$ $2\omega_0$

Fig. S170.2

S171 When the switch is closed, currents, as shown in Fig. S171.1, flow round the circuit.

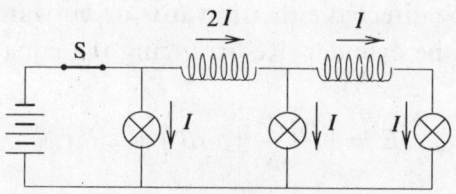

Fig. S171.1

In the period immediately after opening the switch, the current flowing in each coil is practically unchanged; if this were not the case, there would be a rapid change in its magnetic flux which would induce a very high voltage in the coil. Currents of $2I$ and I therefore continue to flow in the coils, and these determine the currents flowing through the lamps (*see* Fig. S171.2).

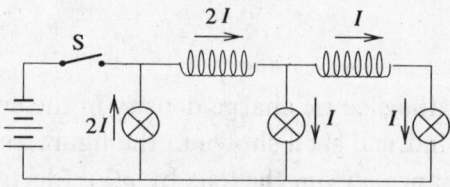

Fig. S171.2

This means that the lamp closest to the switch suddenly flashes, but the brightness of the two other lamps does not change. This all takes place in a very brief period and later all three lamps fade and go out.

S172   The cross-sectional area of the space to be filled is fixed, whilst that of the wire varies as d^2. Thus $n \propto d^{-2}$. The resistance of one turn is inversely proportional to the cross-sectional area of the wire, i.e. varies as d^{-2}, and hence the resistance per unit length of the solenoid is $R \propto nd^{-2} \propto d^{-4}$. The flux density B is $\propto nI$ and therefore the required current $I \propto n^{-1} \propto d^2$. The heat dissipated per unit length is RI^2, which is $\propto d^{-4}\left(d^2\right)^2$, i.e. independent of d. Thus (within limits) it does not matter what diameter wire is chosen so far as the heating effect is concerned.

S173   The equation describing the forces that keep the free electron moving on a circular track (inside the cylinder) is

$$eE \pm er\omega B = mr\omega^2,$$

where e is the charge of the electron, m is its mass, r is its distance from the axis of rotation and E is the strength of the electrostatic field produced in the cylinder by the charge distribution. The $\pm$ sign shows that the Lorentz magnetic force can be directed either inwards or outwards, depending on the sense of rotation of the cylinder. Re-arranging the equation of motion,

$$E = \left(\frac{m\omega^2}{e} \pm \omega B\right) r \equiv Kr,$$

shows the field strength to be directly proportional to the radius.

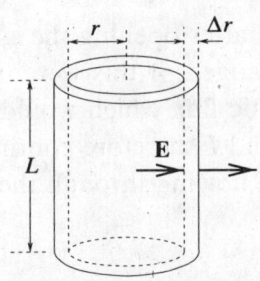

Using Gauss's law, the electric charge density in the cylinder can be found. Consider the thin cylindrical shell shown in the figure and denote the electric charge density at distance r from the axis by $\rho(r)$. Electric flux of magnitude of $2\pi rLE(r)$ enters the shell and a flux of $2\pi(r + \Delta r)LE(r + \Delta r)$ leaves it.

According to Gauss's law

$$K(r + \Delta r)\, 2\pi\, (r + \Delta r)\, L - K r 2\pi r L = (1/\varepsilon_0)\rho\, 2\pi r \Delta r L.$$

From this we obtain

$$\rho = \frac{2\varepsilon_0 \omega m}{e} \left(\omega \pm \frac{eB}{m} \right)$$

for the charge density within the cylinder, noting that it is independent of r.

The density can be positive, negative or zero – depending on the directions and magnitudes of the magnetic field and the angular velocity. It is zero, if $\omega = |e|\, B/m$. For this situation, positive and negative charges are not separated inside the cylinder as the centripetal force provided by the Lorentz effect is just what is needed to sustain circular motion.

> *Note.* In order to get an idea of the orders of magnitude involved, assume that the magnetic field is comparable with the Earth's magnetic field near to the Equator, i.e. $B = 3 \times 10^{-5}$ T. Zero charge density corresponds to an angular velocity $\omega = eB/m = 5.3 \times 10^6\, \text{s}^{-1}$, which is more than 50 million revolutions per minute! Such rapid rotation cannot be realised in practice, since no material could stand it.

S174 The electric charge distribution at any point is the same in the rotating frame $\mathcal{K}'$ as in the laboratory (inertial) frame $\mathcal{K}$, because the charge density is proportional to the number of electrons in unit volume, and both the number and the volume are clearly invariant. It thus follows that the electric charge density is homogeneous in the rotating frame and equal to

$$\rho' = \rho = 2\varepsilon_0 \left(\pm B\omega + \frac{m\omega^2}{e} \right) \approx \pm 2\varepsilon_0 B\omega.$$

The force $\mathbf{F}$ acting on a charged particle must be the same in both frames of reference (as demonstrated by the fact that the elongation of a spring which measures the force is independent of the frame of reference). Thus $\mathbf{F}' = \mathbf{F}$.

In the frame rotating with the cylinder the free electrons of the metal are at rest, and thus the net force exerted on them must be zero (otherwise they would move); if the centrifugal force is neglected this implies that the electric field must also be zero ($\mathbf{E}' = 0$). Further,

$$Q(\mathbf{E} + \mathbf{v} \times \mathbf{B}) = \mathbf{F} = \mathbf{F}' = Q[\mathbf{E}' + (\mathbf{v} + \mathbf{v}_{\text{rel}}) \times \mathbf{B}'],$$

where $\mathbf{v}_{\text{rel}}$ is the relative velocity of the two frames, which is different at different points in the cylinder. Recalling that $E = \omega B r$ and noting that $\mathbf{v}_{\text{rel}} \times \mathbf{B}'$ is directed radially with magnitude $\omega B' r$, we conclude that the magnetic field is (vectorially) the same in both frames, i.e. $\mathbf{B}' = \mathbf{B}$. It follows that in a rotating frame of reference electric *charges* can exist *without* an associated electric field, as illustrated in Fig. S174.1.

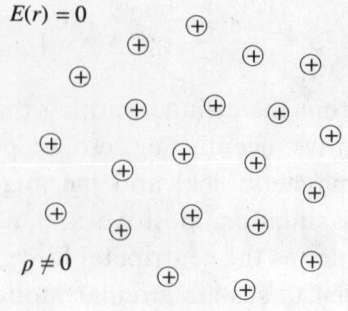

Fig. S174.1

Similarly, it can be proved that in the frame of reference of an observer spinning in a homogeneous magnetic field, an inhomogeneous *electric* field exists, although there are *no* electric charges present (Fig. S174.2).

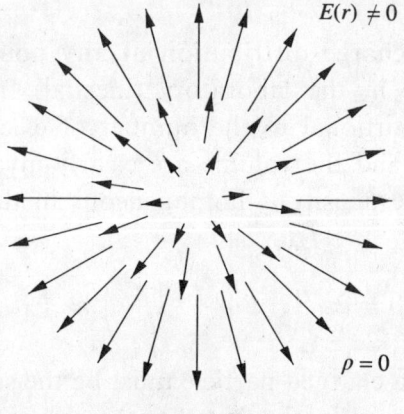

Fig. S174.2

This means that Gauss's law (the connection between electric flux and the charges responsible for it) is *not* obeyed in rotating frames of reference. This surprising result has to be allowed for, even in low-speed (non-relativistic) motion.

S175 It is a mistake to consider this question as a one-dimensional problem. The magnitude of the induced electric field calculated by Jack is correct $[E(r) = rB\omega]$, but the electric field vector within the rotating spoke is not parallel to it; it is radial as shown in Fig. S175.1. This means that the electric flux is non-zero over the curved surface of the cylindrical rod.

Moreover, as shown below, for an elementary cylinder the total flux across this surface is equal in magnitude to the flux across its outer end. It is perhaps surprising that half of the electric flux escapes through the curved surface of the cylinder.

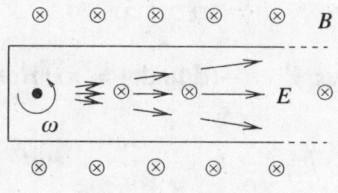

Fig. S175.1

The correct result can be calculated using Gauss's law. The appropriate Gaussian surface is shown in Fig. S175.2. The net electric flux is

$$\Psi = E(r + \Delta r)h(r + \Delta r)\theta - E(r)hr\theta$$
$$= B\omega h\theta \left[(r + \Delta r)^2 - r^2\right] \approx 2Bh\omega\theta r\Delta r.$$

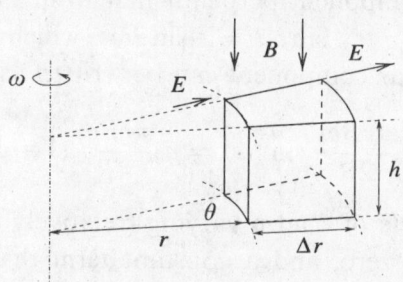

Fig. S175.2

We can relate this to the charge density using

$$\Psi = \frac{1}{\varepsilon_0}\rho\,\Delta V = \frac{\rho hr\theta\,\Delta r}{\varepsilon_0},$$

which gives a value for ρ identical to Jill's.

> *Note.* The sign of the charge density can be either positive or negative, depending on the directions of both the rotation and the magnetic field. The charge distribution inside the spoke is homogeneous, but the net charge on the spoke is zero as there are charges of opposite sign on the spoke's surface. The surface distribution is complicated and cannot be found by elementary methods.

S176 Resolve the magnetic field of the Earth into its horizontal and vertical components. The vertical component induces no current in the ring, since its flux through the ring is always zero. Let the horizontal component of the magnetic field be **B** and the angular velocity of the ring be ω. The magnetic flux threading the ring is

$$\Phi = \pi r^2 B \cos \omega t,$$

and the induced voltage is $V = -(d\Phi/dt) = \pi r^2 B \omega \sin \omega t$. The current,

$$I = \frac{V}{R} = \frac{r^2 \pi B \omega}{R} \sin \omega t,$$

flowing in the ring induces a magnetic field at the centre of the ring of magnitude

$$B_I = \mu_0 \frac{I}{2r} = \mu_0 \frac{\pi r B \omega}{2R} \sin \omega t.$$

The direction of the magnetic field B_I is perpendicular to the plane of the ring and rotates with it. Resolve the vector $\mathbf{B}_I$ into a component parallel to **B** and a component perpendicular to it. The parallel component is proportional to $\cos \omega t \times \sin \omega t = \frac{1}{2} \sin 2\omega t$, which averages to zero over time. The perpendicular component can be written as

$$B_\perp = \mu_0 \frac{\pi r B \omega}{2R} \sin^2 \omega t = \mu_0 \frac{\pi r B \omega}{4R} (1 - \cos 2\omega t).$$

This expression consists of a term varying (relatively rapidly) with time and which, on average, is zero, and a constant term that causes the magnetic needle to deviate by $\alpha = 2°$ from its original (north–south) direction. Since the needle aligns itself with the direction of the (average) resultant field,

$$\tan \alpha = \frac{B_\perp}{B} = \mu_0 \frac{\pi r \omega}{4R}.$$

The needle will make small oscillations about the above position, with an amplitude determined by the mechanical and magnetic characteristics of the needle and by the damping forces.

It is interesting to note that the angle of deviation of the magnetic needle does not depend on the magnitude of the Earth's magnetic field. The only important thing is that its horizontal component is *non-zero*. The resistance of the ring can be calculated from the above formula and is found to be $1.78 \times 10^{-4} \, \Omega$.

S177 With I and i as defined in the hint, the required voltmeter reading is given by RI. In both cases, applying Kirchhoff's laws yields the equations:

$$RI + \frac{\theta}{2\pi}r(I + i) = \frac{1}{2}a^2\dot{B}\lambda,$$

$$ri + \frac{\theta}{2\pi}rI = \pi a^2\dot{B}.$$

In case (a), $\lambda = \theta$ and solution of the simultaneous equations shows that I, and hence the voltmeter reading, is zero.

In case (b), $\lambda = \theta - \sin\theta$ and straightforward but slightly lengthy algebra shows that the voltmeter reading will be

$$|V| = \frac{2\pi^2 Ra^2\dot{B}\sin\theta}{4\pi^2 R + r\theta(2\pi - \theta)}.$$

S178 Consider two touching discs each of radius $R = L/2\pi$ as shown in Fig. S178.1.

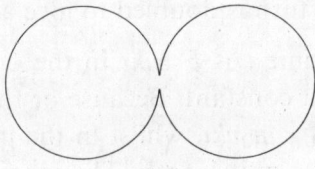

Fig. S178.1

If the discs are placed with their plane perpendicular to a homogeneous magnetic field whose strength changes uniformly with time, the voltage induced in the piece of wire wrapped around their edge is

$$V = 2\pi R^2 \frac{\Delta B}{\Delta t}.$$

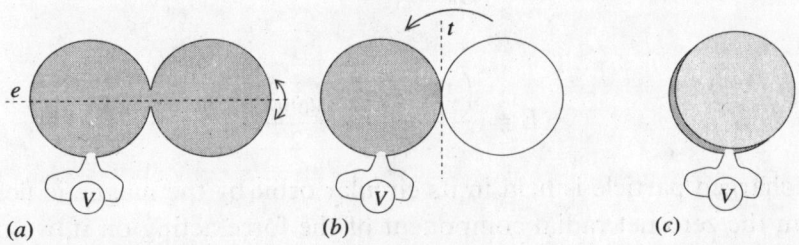

Fig. S178.2

Now twist the disc on the right by 180° about its symmetry axis e (Fig. S178.2(a)). Its top (dark) side then becomes its bottom side (Fig. S178.2(b)). Turn the same disc again by 180°, but this time about axis t (Fig. S178.2(c)). At the end of this, the dark sides of both discs are on top and the perimeter is exactly the same as that of the Moebius strip mentioned in the problem.

Thus, with the strip in a uniformly changing magnetic field, the voltmeter reads

$$V = 2\pi R^2 \frac{\Delta B}{\Delta t} = \frac{kL^2}{2\pi}.$$

This value is much higher than what one would naïvely expect, if reasoning from the area of the paper band. The area of the (one-sided) surface covering the Moebius strip is *not* the same as the area of the paper band, and for narrow strips it is, in fact, much larger!

The induced voltage can also be calculated by cutting the wire at the 'twist' into two 'coils' of one turn each, and adding up the algebraic values of the voltages induced in each turn (taking account of their directions). In the present case, the directions of the two turns are the same, and therefore the voltage $V_0 = k\pi R^2$ in one turn is doubled to give a total voltage of $V = 2V_0$.

S179 The current at time t is $I = kt$ in the outer coil, and $2I = 2kt$ in the inner one, where k is a constant. Because of these currents the magnetic field in the outer coil is $B = \mu_0 nkt$, whilst in the inner one it is $3B$, where n is the number of turns per unit length. The magnetic flux enclosed by the particle's trajectory of radius r is

$$\Phi = \pi R^2 \times 2B + \pi r^2 \times B = \left(2R^2 + r^2\right)\pi\mu_0 nkt.$$

The (constant) magnitude of the induced electric field E can be calculated from the rate of change of magnetic flux with time:

$$E \times 2\pi r = \frac{d\Phi}{dt} = \left(2R^2 + r^2\right)\pi\mu_0 nk,$$

and so

$$E = \frac{\left(2R^2 + r^2\right)}{r}\frac{\mu_0 nk}{2}.$$

The charged particle is held in its circular orbit by the magnetic field, and so, from the zero net radial component of the force acting on it, we obtain

$$\frac{mv^2}{r} = qvB. \tag{1}$$

The particle is accelerated along its circular orbit by the tangential component of the net force according to $ma_t = qE$, where m is the mass and q the electric charge of the particle.

As the magnitude of the electric field is constant, the speed of the particle increases uniformly with time,

$$v = a_t t = \frac{qE}{m}t = \frac{(2R^2 + r^2)}{r} \frac{\mu_0 nk}{2} \frac{q}{m}t.$$

Inserting this and the value of B into equation (1), we get

$$\frac{m}{r}\frac{(2R^2 + r^2)}{r}\frac{\mu_0 nk}{2}\frac{q}{m}t = q\mu_0 nkt,$$

which is satisfied if

$$\frac{(2R^2 + r^2)}{2r^2} = 1, \quad \text{i.e. } r = \sqrt{2}R.$$

S180 The changing magnetic field induces an electric field in the ring. Let us imagine the ring divided into small sections each of length Δs and denote the tangential component of the induced electric field by E_t (in the general case E_t can vary from point to point). The charge on a small section of the ring is

$$\Delta Q = Q\frac{\Delta s}{2\pi r},$$

where r is the radius of the ring. The force exerted on it is

$$\Delta F_t = \Delta Q\, E_t$$

and the resultant torque is

$$\Delta \tau = r\, \Delta F_t.$$

The total torque experienced by the ring is thus

$$\tau = \sum \Delta \tau = \sum r Q\frac{\Delta s}{2\pi r} E_t = \frac{Q}{2\pi}\sum E_t\, \Delta s.$$

Identifying the expression $\sum E_t\, \Delta s$ as the induced electromotive force along the ring, which is directly proportional to the rate of change in the magnetic flux, we have

$$\sum E_t\, \Delta s = -\frac{\Delta \Phi}{\Delta t} = -\pi r^2\frac{\Delta B}{\Delta t}.$$

As a result of the torque, the ring, which has a moment of inertia $I = mr^2$, starts to spin with angular acceleration α. During a time interval Δt its angular velocity changes by

$$\Delta \omega = \alpha\, \Delta t = \frac{\tau}{I}\Delta t = \frac{Q}{2\pi}\left(-\pi r^2\frac{\Delta B}{\Delta t}\right)\frac{1}{mr^2}\Delta t = -\frac{Q}{2m}\Delta B.$$

Since the magnetic field strength increases from zero to B, the final angular velocity of the ring will be

$$\omega = -\frac{QB}{2m}.$$

Note: (i) The negative sign shows that the direction of the angular velocity vector is opposite to the magnetic induction if Q is positive.

(ii) It is interesting to note that the final angular velocity does not depend on the radius of the ring, the time over which the magnetic flux changes, or even on how the magnetic flux increases with time.

(iii) In our calculation we ignored the magnetic field produced by the rotating ring.

(iv) Except in the case of a *cylindrically symmetric* uniform field, it is not possible to find the actual value of the induced electric field within the ring because the geometrical structure of the magnetic field is unknown and we do not know the position of the ring in the magnetic field. We can determine the *total* induced electromotive force, but not the electric field itself.

S181 The resultant magnetic field B' is the sum of the magnetic fields of the Earth and the coil, B_0 and B respectively, i.e.

$$B' = B_0 \pm B. \tag{1}$$

The current flowing through the coil is determined by the induced voltage V_i and the resistance R,

$$I = \frac{V_i}{R} = B' \frac{r^2\omega}{2} \frac{1}{R}, \tag{2}$$

where the induced voltage has been calculated from the rate at which a disc radius cuts the field's magnetic flux. The magnetic field produced by the coil itself is

$$B = \mu_0 nI. \tag{3}$$

From the three equations above, B, B' and I can easily be determined. The two signs occurring in equation (1) allow for both positive and negative values of the angular frequency. The value of ω is taken to be positive if the magnetic field of the coil acts in the same direction as that of the Earth. The following results are obtained for the resultant magnetic field and the current:

$$B' = \frac{2RB_0}{2R - \mu_0 nr^2\omega}, \quad \text{and} \quad I = \frac{B_0 r^2\omega}{2R - \mu_0 nr^2\omega}.$$

As expected, when the disc is at rest, the current is zero and the resultant magnetic field inside the coil is simply B_0, the magnetic field of the Earth.

When the direction of rotation is such that the field in the coil opposes the external magnetic field ($\omega < 0$) the resultant decreases asymptotically to zero as the (negative) angular frequency of the disc increases. At such high rotation speeds, the current flowing in the coil tends to $-B_0/\mu_0 n$ (the value needed to cancel the magnetic field of the Earth).

Rotation of the disc in the opposite direction, ($\omega > 0$), causes the resultant magnetic field to increase. This leads to a higher voltage being induced and a larger current flowing, which in turn leads to a further increase in the magnetic field. Under these positive feedback conditions, both the magnetic field and the current tend to infinity as a particular 'critical' angular frequency, $\omega_{\text{crit}} = 2R/(\mu_0 n r^2)$, is approached, as shown in Fig. S181.1. Such a state is obviously not realised in practice. If the angular frequency is increased too much, the current and the heat given out by the coil increase until the wires burn away!

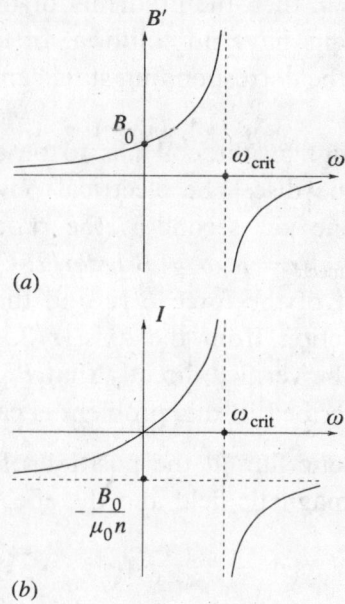

Fig. S181.1

The strange behaviour of the system can be more easily understood if the relationship between the current in the coil and the resultant magnetic field is represented graphically as in Fig. S181.2. According to equation (2), I is proportional to B' in such a way that the coefficient of proportionality depends on ω.

This is represented by a straight line through the origin, with a gradient proportional to ω. Equations (1) and (3) show that $B' = B_0 + \mu_0 n I$; this is also a linear relationship, but its graph does not pass through the origin.

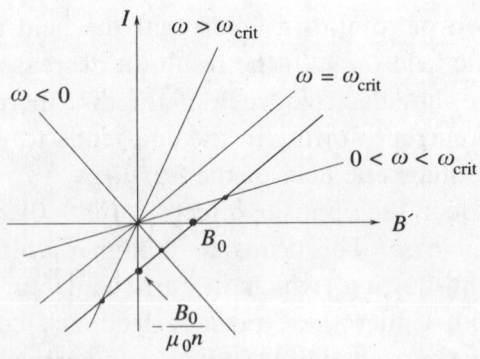

Fig. S181.2

The gradient of the latter is $1/(\mu_0 n)$ and independent of ω. The intersection of these two straight lines determines the actual current and the resultant magnetic field. If $\omega = \omega_{\text{crit}}$, then the gradients of the two straight lines are the same and the equations have no solution. In fact, the critical angular frequency is so high that the corresponding state cannot even be approached in practice.

The Joule heat given out by the coil has to be equal to the mechanical work done in rotating the disc. The electrical power is $P_{\text{el}} = I^2 R$, while the mechanical work done per second is the product of the torque and the angular frequency, $P_{\text{mech}} = M\omega = B'Ir^2\omega/2$. (The torque M has been calculated as the product of the force $B'Ir$ and the average perpendicular distance of its line of action from the axis, $r/2$.) Using the relationship between B' and I, it can be verified directly that $P_{\text{el}} = P_{\text{mech}}$.

The strange device described in this problem is called a *unipolar dynamo*.

S182   The total magnetic flux at the position of the ring is made up of that due to the external magnetic field and the effects of self-induction,

$$\Phi = B_z \, \pi r_0^2 + LI.$$

Any change in magnetic flux induces a current in the ring in accordance with

$$RI = \frac{\Delta\Phi}{\Delta t}.$$

However, this has to be zero since the ohmic resistance of a superconducting ring is zero. Accordingly, the magnetic flux through the ring has to be constant, i.e.

$$\Phi = B_0(1 - \alpha z) \, \pi r_0^2 + LI = \text{constant.}$$

From the initial conditions ($z = 0$, $I = 0$), the value of the constant is $\Phi = B_0 \pi r_0^2$.

The current in the ring can be determined using the above equations which give

$$I = \frac{1}{L} B_0 \alpha \pi r_0^2 z.$$

The Lorentz force acting on the ring (which can only be vertical, because of the symmetry of the assembly) can be expressed as

$$F_z = -B_r I(z) 2\pi r_0 = -B_0 \beta r_0 \frac{1}{L} B_0 \alpha \pi r_0^2 2\pi r_0 z = -kz.$$

The Lorentz force is thus directly proportional to the vertical displacement of the ring, with the coefficient of proportionality calculable from the given data. (This result is only valid for small displacements, since the magnetic induction is not adequately described by the given formulae for large ones.)

The equation of motion of the ring is

$$ma_z = F_z - mg = -kz - mg.$$

This means that the ring makes harmonic oscillations about the equilibrium position $z_0 = -mg/k$ with

$$z(t) - z_0 = A \cos \omega t,$$

where $\omega = \sqrt{k/m}$. From the initial conditions it follows that $A = -z_0$, and so

$$z(t) = \frac{g}{\omega^2} (\cos \omega t - 1).$$

The vertical z-coordinate is never positive, and it follows that the Lorentz force always points upwards, being zero at the topmost point of the oscillation. The current always flows in the same direction around the ring.

Substituting the numerical data gives $\omega = 31.2 \text{ s}^{-1}$ and $A = 1$ cm. The time dependence of the current flowing in the ring can be expressed in terms of $z(t)$ as

$$I = \frac{1}{L} B_0 \alpha \pi r_0^2 z(t) = \frac{1}{L} B_0 \alpha \pi r_0^2 A (\cos \omega t - 1).$$

The maximum value of the current, which flows at the bottom of the oscillation, is $I_{max} = 39$ A.

S183   Let us apply the law of conservation of energy for a particle of mass m and charge Q:

$$\frac{1}{2}mv^2 + QK\frac{\cos\theta}{r^2} = \frac{1}{2}mv_0^2 + \frac{QK\cos(\pi/2)}{r^2} = 0.$$

We can then express the velocity of the bead at angle θ as

$$v = \sqrt{-2\frac{QK}{mr^2}\cos\theta}, \qquad (\tfrac{\pi}{2} \le \theta \le \pi).$$

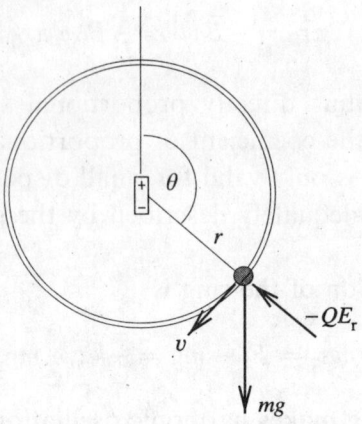

The circular motion needs a radial force component of mv^2/r. The radial component of the force on a unit charge due to the dipole (i.e. the effect of the radial component of the electric field) can be calculated as minus the derivative of the electric potential with respect to r,

$$E_r = -\frac{\partial\Phi}{\partial r} = 2\frac{K\cos\theta}{r^3}.$$

Using the earlier expression for the velocity, we notice that QE_r is just equal to $-mv^2/r$, the required centripetal force. Thus the string does not need to exert any force on the bead to sustain circular motion. If the string were not there, the bead would move along a circular path until it reached the point opposite its starting position. The bead would stop there, and then repeatedly retrace its path executing a periodic motion.

> *Note.* The time dependence of this motion is just the same as that of a simple pendulum subject to gravity after release from a displacement of 90°.

S184   Imagine that you are sitting in a frame of reference moving horizontally at constant speed v perpendicular to the magnetic field lines, and you are carrying a charge q. In a frame of reference fixed to the ground, the charge moving with velocity v is observed to experience a force of magnitude qvB (pointing upward or downward, depending on the direction of the motion).

In the moving coordinate system, the charge is at rest and the Lorentz force does not act on it, though it still 'feels' the force. (The presence of a force caused by some interaction cannot depend on the coordinate system in which it is described.) This seeming contradiction can be resolved by transforming the electric and magnetic fields when changing from a stationary coordinate system to a moving one. In the present case, there is only a magnetic field in the stationary frame of reference and no electric field. In a frame that moves (at a speed much less than the speed of light), this same magnetic field can be observed, but an electric field of field strength $E = vB$ is also present. This provides the force $F = qE = qvB$ that would otherwise be missing in this frame.

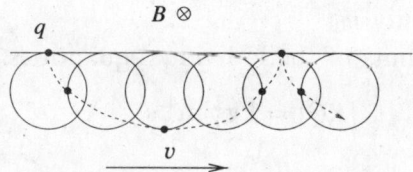

Let the velocity of the moving frame of reference be such that the electric force described above is the same as the gravitational force mentioned in the problem, i.e.

$$qvB = qE = F = mg.$$

We now describe the motion of the observed body in this coordinate system. Since the effects of the electric and gravitational fields cancel each other out and the body moves at velocity $-v$ in this system, a force of $-qvB$ (the magnetic Lorentz force) acts on the body. This makes it move on a circular track spiralling downward according to $qvB = mv^2/r$, where r is the radius of the circle and is given by $r = mv/qB$. Using the values of the velocity and the radius, the time in which the body makes one orbit is found to be $T = 2\pi m/qB$, independent of the velocity v.

In the coordinate system fixed to the ground, a uniform rectilinear motion is superimposed on this uniform circular motion. The particle therefore follows a cycloidal path as shown in the figure, falling down to $2r = 2mv/qB$ and then rising again to reach its initial height in time T. The corresponding horizontal displacement is $Tv = 2\pi r$. At this point the particle stops for a moment, before starting out on a new cycloid curve.

S185 When the magnet is falling with the constant terminal speed and it covers a distance L ($L \gg h$), its potential energy loss mgL is converted to Joule heat by the currents induced in L/h loops. Denoting by Q the heat produced in one particular loop, we can write:

$$mgL = \frac{L}{h}Q, \qquad \text{i.e. } Q = mgh.$$

On what quantities can Q depend? Since the power dissipated in a resistor by a given voltage is inversely proportional to its resistance, $Q \propto R^{-1}$, when the magnet is moving with velocity v_0 we can express the heat function in the form

$$Q = \frac{1}{R} f(v_0, \mu, r).$$

Thus it is a general function of the terminal velocity v_0, the magnetic moment μ and the radius r of the circular loop; moreover the formula could involve the vacuum permeability μ_0.

Now consider the units of the individual quantities. The units of RQ are

$$[RQ] = \text{kg}^2 \ \text{m}^4 \ \text{s}^{-5} \ \text{A}^{-2},$$

whilst

$$[v_0] = \text{m s}^{-1}, \qquad [\mu] = \text{A m}^2, \qquad [r] = \text{m}, \qquad [\mu_0] = \text{kg m s}^{-2} \ \text{A}^{-2}.$$

Thus it must be that

$$RQ(v_0, \mu, r, \mu_0) \propto v_0 \ (\mu_0 \mu)^2 \ r^{-3}.$$

But we know that $Q = mgh$, and so

$$v_0 \propto \frac{mghRr^3}{\mu^2}.$$

It follows that the terminal speed ratios in question are: 2, $\frac{1}{4}$, 2, 2 and 8, respectively.

S186 The speed of the electrons remains constant in the frame of reference fixed to the vacuum chamber (S) because the magnetic field can only change the direction of the velocity of the moving charge. Consider another frame of reference (S') moving parallel to the wire with constant speed v_0 with respect to the first one. We can write the Lorentz force experienced by a particle of charge Q in either frame:

$$\mathbf{F}' = \mathbf{F} = Q(\mathbf{v} \times \mathbf{B}) = Q(\mathbf{v}' + \mathbf{v}_0) \times \mathbf{B} = Q(\mathbf{v}' \times \mathbf{B}') + Q\mathbf{E}'.$$

We first note that the electric current in both frames of reference must be the same. In frame S, there are moving electrons and static positive ions whilst the wire itself is neutral. In reference frame S', the speed of electrons and so the current of electrons will be different, but the current of the moving positive ions just compensates for this change. The consequence of this is that there is the same magnetic field in both frames ($\mathbf{B}' = \mathbf{B}$). The equation above shows that in frame S' we have, in addition to the unaltered magnetic field, an electric field given by $\mathbf{v}_0 \times \mathbf{B}$. (Note that the transformation of a single magnetic field from S to S' does not generally alter the magnetic field if the speed of the moving system is much less than the speed of light.)

Let us now describe an electron's motion in reference frame S'. In this frame there is an electric field (perpendicular to the wire) of strength

$$E(r) = v_0 B(r) = \frac{\mu_0 v_0 I}{2\pi} \frac{1}{r},$$

where r is the distance from the wire. Using the analogy of the cylindrical capacitor, we can find an appropriate electric potential function for this field (i.e. one that is such that minus its derivative with respect to r gives the field), namely,

$$U(r) = -\frac{\mu_0 v_0 I}{2\pi} \ln r.$$

In the frame S', the initial speed of the electron (at a distance r from the wire) is $\sqrt{2}v_0$, and the electron just stops when it has approached within $r_0/2$ of the wire. We can apply the work–energy theorem to the motion of the electron as follows,

$$\frac{1}{2} m \left(\sqrt{2} v_0 \right)^2 = -\frac{\mu_0 v_0 I Q}{2\pi} \left(\ln \frac{r_0}{2} - \ln r_0 \right) = \frac{\mu_0 v_0 I Q}{2\pi} \ln 2.$$

Inserting numerical values into this expression, gives the initial speed of the electron as

$$v_0 = \frac{\mu_0 Q}{2m\pi} I \times \ln 2 \approx 2.46 \times 10^5 \text{ m s}^{-1} \approx 250 \text{ km s}^{-1}.$$

Note. (i) The initial speed of the electron is very large on a macroscopic scale, but is very small for electrons; an electron attains this speed if it is accelerated through a potential difference of as little as 0.2 V. Since 250 km s^{-1} is much less than c, we justifiably ignored the relativistic mass increase of the electron in this problem.

(ii) It is interesting to observe that if the electron cannot approach the wire closer than $r_0/2$, then the maximum distance of the electron is also limited – it cannot be further away than $2r_0$. More generally, if the minimum distance is r_0/n, then the maximum distance must be nr_0. This can be proved using another frame of reference moving with velocity $-v_0$.

S187   The Lorentz force

$$F = Q(E + v \times B)$$

acts on a small body of charge Q moving with velocity v in a field of electric field strength E and magnetic field strength B. In a frame of reference moving at velocity v_0 relative to the original one, the velocity of the particle is $v' = v - v_0$. If all physical quantities in this new reference frame are denoted by primes, the expression for its Lorentz force is

$$F' = Q'(E' + v' \times B').$$

Now compare the two frames of reference. Any force acting on the particle which can be detected (e.g. through the acceleration it produces) cannot change, i.e. $F' = F$. The same is true for the electric charge: $Q = Q'$. (If particle charges depended on their velocity then an initially neutral body would show a net electric charge when heated – no trace of this is seen in nature.)

Transforming the velocities yields

$$Q(E + v \times B) = Q(E' - v_0 \times B' + v \times B').$$

This relation has to be satisfied for any v, and specifically for stationary particles. Therefore

$$B' = B \qquad \text{leading to} \qquad E' = E + v_0 \times B.$$

It can be seen that electric and magnetic fields are not independent physical quantities, their values depend on the (velocity of the) frame of reference in which phenomena are described (*see also* P184). We now apply these general relationships (twice) to the problem in hand.

Consider the frame of reference of the liquid. To get to this frame from the capacitor's frame requires a transformation with $v_0 = -v$. The magnetic field in this new frame is unaltered as $B' = B$, but an electric field E' of strength $-v \times B$ is also present. The stationary liquid is polarised by this field, and the electric field strength is consequently reduced by a factor $1/\varepsilon_r$ to $E' = -v \times B\,(1/\varepsilon_r)$. Returning to the original (capacitor) frame of reference requires a further transformation, but this time with v_0 set equal to $+v$. Again the magnetic field is unchanged, $B'' = B' = B$, but the electric field is given by

$$E'' = E' + v \times B' = -v \times B\,(1/\varepsilon_r) + v \times B = \left(1 - \frac{1}{\varepsilon_r}\right) v \times B.$$

Consequently, a p.d. of

$$V = vBd \left(1 - \frac{1}{\varepsilon_{\mathrm{r}}}\right)$$

appears between the plates.

For non-polarisable materials ($\varepsilon_{\mathrm{r}} = 1$), this voltage is obviously zero, whilst in the case of 'easily' polarisable materials (metals), $V = vBd$. This latter equation, describing the Hall effect, can be interpreted in the following way. The charges in the conductor moving in the magnetic field are displaced by the Lorentz force. The sideways migration of charges continues until an electrostatic field strong enough to balance the Lorentz force (which is proportional to the magnetic field) has been built up. From the condition for this, $QE = QvB$, the above relationship, $V = Ed = vBd$, can be recovered.

S188 It is not correct in quantum mechanics to describe processes in terms of their various parts occurring in a particular order, nor to describe classically conserved quantities as being 'borrowed'. However, the 'classical' description of a quantum effect usually gives a qualitatively correct picture if the 'borrowing' of a quantity is interpreted as meaning that the more of it that is borrowed, the less likely the process is to occur.

The volume of the uranium nucleus is equal to the sum of the volumes of its fission products, but the total surface area of the fission products is greater than that of the original uranium nucleus. This means there is a temporary loss of (binding) energy when the process starts (this is the energy that has to be 'credited' to initiate the fission process), which is later refunded by the decrease in Coulomb energy of the system when the daughter nuclei move away from each other. If the uranium nucleus were to split into three fission products, the energy required at the start of the process would be so much higher than for two, that, both in theory and in practice, the corresponding probability would be negligible.

S189 Even if atoms are heated to a few thousand degrees, their thermal energy is still much smaller than their binding energy per nucleon. For this reason the rate of nuclear reactions is usually independent of temperature.

However, the isotope ^{7}Be transforms into ^{7}Li via K-capture (the capture of an electron from the innermost K shell). At a temperature of a few thousand degrees the quantum of thermal energy kT is comparable with the ionisation energy of the two innermost electrons of the Be atom (consider the tail of the Maxwell velocity distribution). If some fraction of the beryllium atoms becomes ionised, the probability of nuclear electron capture occurring decreases by approximately the same percentage. Thus, at such temperatures, the radioactivity, and hence the half-life, of beryllium can be affected.

S190   In equilibrium, abundances are proportional to half-lives. In 10^{-3} kg of thorium, almost exclusively ^{232}Th, there will be

$$\frac{6.02 \times 10^{23}}{232} = 2.595 \times 10^{21}$$

atoms of thorium. Consequently, there will be

$$2.595 \times 10^{21} \times \frac{56 \text{ s}}{1.41 \times 10^{10} \text{ y}} \approx 3.3 \times 10^{5}$$

atoms of radon.

Over the stated time span both isotopes of thorium will contribute to the radon population. Each contributes an amount which first increases at a rate governed by the slowest decay rate in the relevant intermediate decay chain, then remains steady at the equilibrium value, and finally decays at a rate governed by the decay rate of the parent.

Thus the contribution from the $^{228}_{90}$Th rises with a half-life of 3.64 days and decays with a half-life of 1.91 years. Similarly, the contribution from the $^{232}_{90}$Th rises with a half-life of 5.7 years and decays with a half-life which is essentially infinite. The total $^{220}_{86}$Rn present is the sum of these two as shown in the figure.

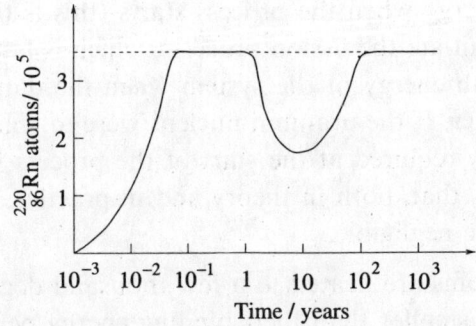

The initial rise in the curve is due to radon produced from the $^{228}_{90}$Th present in the initial sample. Most of this has decayed away by the time that the radon originating from the original $^{232}_{90}$Th becomes significant. After about eight years the numbers of radon atoms from the two different ancestors become equal at about 0.9×10^{5} each. After that the number again rises to the equilibrium value of 3.3×10^{5} with virtually all atoms present having $^{232}_{90}$Th as their ancestor; it would remain at this value for about 10^{10} years.

S191 The relativistic energy and momentum of a particle with (rest) mass m and speed v are

$$E = \frac{mc^2}{\sqrt{1 - v^2/c^2}} \quad \text{and} \quad p = \frac{mv}{\sqrt{1 - v^2/c^2}},$$

where c is the velocity of light. These two equations can also be combined as $E = \sqrt{p^2c^2 + m^2c^4}$.

As a result of particle–particle collisions, new (additional) particles, e.g. proton–antiproton pairs, can be produced. Antiprotons have the same rest mass as protons. The total energy of the final particles is lowest if they stay together, moving with negligible speed (in the reference frame of their centre of mass). For this situation, the four particles (three protons and an antiproton) can be considered as one particle of mass $4m$, whose momentum is equal to the initial momentum of the accelerated proton. The law of conservation of energy prescribes that

$$\sqrt{p^2c^2 + m^2c^4} + mc^2 = \sqrt{(4m)^2c^4 + p^2c^2}.$$

Squaring both sides of the equation gives $p = \sqrt{48}mc$ and $E_p = 7mc^2$. The protons thus have to gain a kinetic energy of $E_p - mc^2 = 6mc^2 \approx 6$ GeV in the course of the acceleration, which requires an accelerating voltage of 6×10^9 V.

S192 Electrons of mass m and charge e can move freely in the wall of the Faraday cage to reach a state in which the resultant of the electric and gravitational forces acting on them is zero. This requires a small condensation of the electrons at the bottom of the metal wall, leading to a surplus of positive charge (lack of electrons) at the top. The charge displacement continues until the magnitude of the resulting vertical electric field is $E_0 = mg/|e|$.

As a homogeneous electric field is formed in the wall of the cage, the same field has to be present inside the cage itself. If this were not the case, the motion of a charged particle vertically upwards or downwards in the wall of the cage, and then back to its initial position through the inner part of the cage, would form the basis of a perpetual motion machine – and obviously this is not possible.

Thus, as a result of charge displacement in the wall, an electrostatic field is set up which acts on a free electron placed inside the cage with the same force as gravity does. In other words, the cage shields the gravitational field as well! This shielding, however, only works for electrons; the condition of zero net force is not satisfied for particles with a different ratio of charge to mass (e/m). In the case of positrons, for example, the resultant force is $E_0|e| + mg = 2mg$, i.e. these particles start moving vertically downwards with an acceleration of $2g$.

Numerically, $E_0 = 5.6 \times 10^{-11}$ V m^{-1}, a very small value, and therefore the phenomenon described cannot be seen under ordinary circumstances. Electrons cannot be placed into a Faraday cage with no initial speed, since even in the case of the photoelectric effect, in which electrons are pushed out of the metal with an energy of only 0.1 eV, they still travel with a speed of 200 km s^{-1}. The relative change in velocity after 1 metre of free fall is only about one part in 10^{10}, which is immeasurably small.

> *Note.* So far, we have not considered the effect of the displacement of the charges in the wall due to the presence of the positron in the cage. The magnitude of this effect can be calculated using the method of image charges. A charge e at a small distance d from the wall, experiences a force caused by the polarised charges in the metal that is the same as the force F_{image} produced by a charge of $-e$ at distance $2d$ from the first charge (an 'image charge'), i.e.
>
> $$F_{\text{image}} = k\frac{(-e)e}{(2d)^2}.$$
>
> In the case of electrons or positrons, this force is comparable to their weight if $d \approx 2$ m. Therefore if somebody wanted to verify the conclusions of the above reasoning using experimental data, the experimenter would need to build a Faraday cage large enough to allow the particles to be more than a few metres away from the wall. If it were smaller than this, it would not be the shielding effect of the cage which was being examined, but the effect of polarising the charges in the walls.

S193   Let m and M denote the masses of the positron and the proton, respectively ($M \approx 2000m$), and let e denote the elementary charge. Because of the large mass ratio, the protons will hardly have moved by the time the positrons are already far away. We equate the energy of the initial state to that of the one in which the positrons have moved much further than 1 cm away and are travelling at speed v_1:

$$\frac{e^2}{4\pi\varepsilon_0}\left(\frac{4}{a} + \frac{2}{\sqrt{2}a}\right) = \frac{e^2}{4\pi\varepsilon_0}\frac{1}{\sqrt{2}a} + 2\frac{1}{2}mv_1^2.$$

Substitution of the numerical data yields $v_1 = 350$ m s^{-1}. Thereafter, the speed of the positrons does not change significantly.

The protons, on the other hand, although hardly moving whilst the positrons are close, do repel each other and are accelerated to a speed

v_2. Conservation of energy can again be applied and now gives

$$\frac{e^2}{4\pi\varepsilon_0}\frac{1}{\sqrt{2}a} = 2\frac{1}{2}Mv_2^2.$$

This yields the value $v_2 \approx 2.7$ m s^{-1}.

S194 The general equations of conservation of energy and momentum for Compton scattering are:

$$hf_0 + m_0c^2 = hf + mc^2,$$

$$\frac{hf_0}{c} = \frac{hf}{c}\cos\alpha + p\cos\beta,$$

$$\frac{hf}{c}\sin\alpha = p\sin\beta,$$

$$\left(mc^2\right)^2 = \left(m_0c^2\right)^2 + p^2c^2,$$

where f_0 and f denote the frequencies of the incident and scattered photon respectively, α and β are the photon and electron scattering angles (*see figure*) and m_0 is the rest mass of the electron.

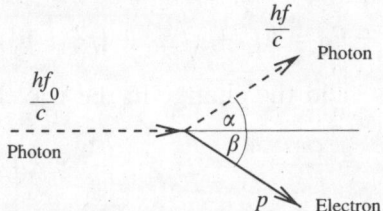

For the special case considered,

$$hf_0 = m_0c^2 \qquad \text{and} \qquad \frac{hf}{c} = p.$$

Solving this system of equations we find that

$$\alpha = \beta = \cos^{-1}\frac{2}{3} \approx 48.2°,$$

and so the angle between the scattered photon and recoil electron is $\alpha + \beta \approx 96.4°$. The momentum of the scattered electron is

$$p = \frac{3}{4}m_0c = \frac{m_0v}{\sqrt{1 - v^2/c^2}},$$

yielding the speed of the electron as $v = \frac{3}{5}c$. This is of the same order of magnitude as c, and consequently it was appropriate to use relativistic formulae.

S195 The energy of a photon of frequency f is hf, while its momentum is hf/c, where h is Planck's constant and c the velocity of light. The energy of the photon decreases to hf' after the collision, and its momentum, of magnitude hf'/c, is perpendicular to the original momentum.

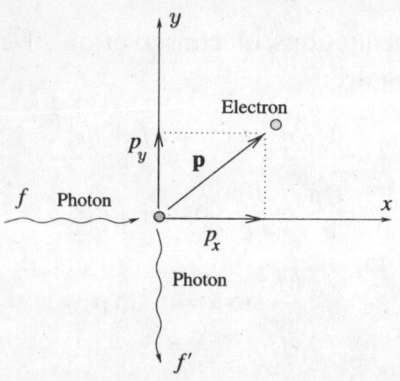

It follows from momentum conservation that, as shown in the figure, the components of the electron's final momentum are $p_x = hf/c$ and $p_y = hf'/c$. Using the principle of conservation of energy in its relativistic form, $E^2 = E_0^2 + p^2c^2$, where E_0 is the rest-mass energy, we obtain

$$hf + E_0 = hf' + \sqrt{E_0^2 + (p_x^2 + p_y^2)c^2}.$$

The above relationships yield the change in the wavelength of the photon as

$$\Delta\lambda = \lambda' - \lambda = \frac{c}{f'} - \frac{c}{f} = \frac{hc}{E_0} = \frac{h}{mc} \approx 2.4 \times 10^{-12} \text{ m}.$$

Note. This interesting quantity with the dimensions of a length – dependent only on the mass of the electron, the speed of light and Planck's constant, i.e. natural constants – is called the Compton wavelength of the electron.

S196 Applying the energy formula for spherical capacitors to the 'classical electron' gives

$$\frac{1}{2}\frac{e^2}{4\pi\varepsilon_0 r} < mc^2,$$

which we can write as

$$r > \frac{1}{4\pi\varepsilon_0}\frac{e^2}{2mc^2} = 1.4 \times 10^{-15} \text{ m}.$$

This value is called the classical electron radius.

The classical electron is considered as a rotating sphere of this radius with a homogeneous mass distribution, an angular frequency ω and an angular momentum corresponding to that of a rotating rigid body. Hence we obtain $\frac{2}{5}mr^2\omega = h/(4\pi)$. This gives the 'equatorial speed' as $v_{eq} = r\omega \approx 350c$. Since this speed is many times higher than the limiting speed of relativistic physics, namely the speed of light, the original version of the classical electron model – worked out by *Lorentz* and *Abraham* before the birth of quantum theory in the early 1900s – had to be abandoned.

A modified version suggested that the electron could be a sphere of radius r, charged on its surface, which did not rotate but would contain a magnetic dipole to account for the experimental fact that electrons have a magnetic momentum. Outside the sphere, a multiple of the vector $\mathbf{E} \times \mathbf{B}$ produced by the electric and magnetic fields, describes the current of electromagnetic momentum. This current carries angular momentum and with a suitable choice of the parameters can be made to equal the measured one.

S197 Assume that the electron occupies a horizontal layer of thickness H just above the bottom of the box. Its vertical coordinate is then known with accuracy H. Therefore the uncertainty in its vertical momentum has to be at least $\Delta p = \hbar/\Delta x = \hbar/H$, where the quantity $\hbar$ is $1/(2\pi)$ times the Planck constant. In such circumstances, the electron has a potential energy of

$$E_{\text{potential}} = mgx_{\text{average}} \approx mg\frac{H}{2},$$

and a kinetic energy of

$$E_{\text{kinetic}} = \frac{\Delta p^2}{2m} \approx \frac{\hbar^2}{2mH^2}.$$

Thus its total energy is

$$E(H) = E_{\text{potential}}(H) + E_{\text{kinetic}}(H) = AH + B\frac{1}{H^2},$$

where A and B are constants determined by the above equations. If H is small, the potential energy is low but the kinetic energy is high. If, on the other hand, H is large, the kinetic energy is low but the potential energy is high. The total energy will be at a minimum if $E_{\text{potential}}$ and E_{kinetic} are of the same order of magnitude. It can be shown using differential calculus that the optimum value for the energy ratio is 2 : 1 in favour of the potential energy. As we only want a rough estimate, the ratio of the energies can be taken to be unity. This gives

$$H \approx \left(\frac{\hbar^2}{m^2g}\right)^{1/3} \approx 1 \text{ mm.}$$

This is quite a large value compared with usual microscopic sizes. The reason for this is that gravitation is very *weak* compared to the electromagnetic interaction which determines the binding energies and sizes of molecules.

S198   If a nucleus of atomic number Z restricts an electron to a sphere of radius r, then the electrostatic energy of the electron is $E_{el} \approx -kZe^2/r$ and its momentum can be estimated to be $\hbar/r$. If $Z \gg 1$ (i.e. the nucleus is a heavy one), the kinetic energy of the strongly bound electron can be calculated using the relativistic formula

$$E_{kin} = \sqrt{p^2c^2 + m^2c^4} - mc^2 \approx pc,$$

where c is the velocity of light and $\hbar$ is the Planck constant divided by 2π. The total energy of the electron is

$$E(r) = E_{el}(r) + E_{kin}(r) = -\frac{1}{4\pi\varepsilon_0}\frac{Ze^2}{r} + \frac{\hbar c}{r}.$$

Since the fine-structure constant $e^2/(4\pi\varepsilon_0\hbar c)$ is approximately $1/137$, the total energy of the electron for small values of r is

$$E(r) = \frac{e^2}{4\pi\varepsilon_0}(137 - Z)\frac{1}{r}.$$

(The above expression is not valid for large values of r, as a small, non-relativistic momentum would have to be taken into account.)

According to the above formula, the electron would fall into the nucleus if $Z > 137$; or rather, it would be confined to a small volume of nuclear size. The figure 137 is only an estimate of the critical atomic number; more accurate calculations, which take the finite size of the nucleus into account, yield values around 150–160 for Z_{crit}.

These calculations show that an electron could be confined to the nucleus of an element with atomic number greater than 150, if transuranic elements of such high atomic numbers could be made at all.

S199   The speed of propagation of surface water waves depends on the surface tension γ and the density ρ of water, and on their wavelength, λ. The dimensions of these quantities are

$$[\gamma] = \frac{N}{m} = \frac{kg}{s^2}, \qquad [\rho] = \frac{kg}{m^3}, \qquad [\lambda] = m.$$

An expression for velocity can only be derived from these quantities if γ and ρ, appear in the combination γ/ρ (otherwise the velocity, which involves only length and time, would have to depend upon the unit of mass). However, as

$$\left[\frac{\gamma}{\rho}\right] = \frac{m^3}{s^2},$$

this expression has to be further divided by the wavelength, and then square-rooted in order to make the result have the dimensions of speed. In summary, dimensional analysis dictates that the speed of propagation of capillary waves is proportional to the reciprocal of the square-root of the wavelength,

$$v \sim \sqrt{\frac{\gamma}{\rho\lambda}} \sim \frac{1}{\sqrt{\lambda}}.$$

From this functional dependence (and the given data), we can conclude that the speed of propagation of surface waves would reach that of sound in water when their wavelengths are of the order of 10^{-8} cm.

Since the speed of propagation of surface waves cannot be greater than that of sound (the molecules cannot transmit a disturbance to each other faster at the surface than inside the matter), waves of wavelength less than approximately 10^{-8} cm have no meaning. This is, in fact, the order of magnitude of the size of water molecules!

S200 You will have noticed not only the acceleration, but also the increasing size of the champagne bubbles. As the sparkling champagne is super-saturated with carbon dioxide, gas is released continuously whilst the bubbles are rising. This is why the size of the bubbles increases, as does the buoyancy force provided by the liquid. The upthrust is proportional to the volume of the bubble, while the viscous drag, which is also increasing, is only proportional to the surface area of the bubble. Consequently, the net upward force increases with bubble size. However, increasing speed also leads to a larger viscous drag, and ultimately the bubble moves under the influence of a collection of balanced forces.